Ethics and Teaching

EXHIBIT "A"
"Buckets of Rain:"
Life is sad
Life is a bust
All ya can do is do what you must.
You do what you must do and ya do it well

"Silvio:"
"I can tell you fancy, I can tell you plain
You give something up for everything you gain
Since every pleasure's got an edge of pain
Pay for your ticket and don't complain."

"Mississippi:"
"Everybody movin' if they ain't already there
Everybody got to move somewhere."
"You can always come back, but you can't come back all the way."

"Tomorrow is a Long Time:"
"[C]an't even remember the sound of my own name."

"Trying to Get to Heaven"
"I been to Sugar Town, I shook the sugar down
Now I'm trying to get to heaven before they close the door."

"Nettie Moore"
"If I don't do anybody any harm, I might make it back home alive."

"It's Alright Ma (I'm Only Bleeding)"
"I got nothing, Ma, to live up to."

"When the Deal Goes Down:"
"We live and we die, we know not why/But I'll be with you when the
deal goes down."

"Sugar Baby:"
"You always got to be prepared but you never know for what."

"The Weight"
"Take a load off Fanny, take a load for free."

Ethics and Teaching

A Religious Perspective on Revitalizing Education

ALAN A. BLOCK

palgrave
macmillan

ETHICS AND TEACHING
Copyright © Alan A. Block, 2009.

First published in 2009 by
PALGRAVE MACMILLAN®
in the United States—a division of St. Martin's Press LLC,
175 Fifth Avenue, New York, NY 10010.

Where this book is distributed in the UK, Europe and the rest of the world,
this is by Palgrave Macmillan, a division of Macmillan Publishers Limited,
registered in England, company number 785998, of Houndmills,
Basingstoke, Hampshire RG21 6XS.

Palgrave Macmillan is the global academic imprint of the above companies
and has companies and representatives throughout the world.

Palgrave® and Macmillan® are registered trademarks in the United States,
the United Kingdom, Europe and other countries.

ISBN-13: 978–0–230–61239–6
ISBN-10: 0–230–61239–3

Library of Congress Cataloging-in-Publication Data

Block, Alan A., 1947–
    Ethics and teaching : a religious perspective on revitalizing education /
    by Alan A. Block.
        p. cm.
    ISBN 0–230–61239–3
    1. Block, Alan A., 1947– 2. Teachers—United States—Biography.
3. Teaching—United States—Philosophy. 4. Jewish ethics. I. Title.

LA2317.B5446B58 2009
371.102—dc22   2008032840

A catalogue record of the book is available from the British Library.

Design by Newgen Imaging Systems (P) Ltd., Chennai, India.

First edition: March 2009

10 9 8 7 6 5 4 3 2 1

Printed in the United States of America.

To Bill Pinar,
for his enduring friendship

Contents

Hakdamah*

One summer my wife and I enrolled our two children at Suzuki Camp, held every year at the University of Wisconsin-Stevens Point. This is a total experience, a five-day immersion in a specific pedagogy of music: when the students (mostly children but a few adults) are not in private lesson they are engaged in performing in groups, an integral element in the Suzuki method. Furthermore, every evening there is a musical event, sometimes a recital by an accomplished artist who can (and often will) attest to the efficacy of the Suzuki method. Also scheduled as part of nightly events are group and solo performances by camp attendees. The days and nights are filled with the sound of music and its approximations. And when campers are not attending a scheduled event, they are encouraged to practice, practice, practice! Our children's violin teacher, who was also a lead instructor at this camp, would say, "Practice only on those days on which you eat." We practiced, and we ate, and then we practiced some more.

During the day, everyone was encouraged to shop at the Suzuki center display. Arrayed brilliantly atop display desks, everything a musician might need immediately or in the future was available for purchase. There were items for the instruments themselves, such as strings, and bridges, and expensive new bows; there were cute little learning devices to enhance and brighten practices for younger players; and there was paraphernalia for every violinist—instrument cases, and backpacks, and music cases and stands. Behind desks sat technicians who could expertly and quickly rehair balding bows, or execute minor repair on injured instruments. There were also items for sale for which I could not discern a purpose, although someone must have deemed them essential. And there, on one small but prominently situated table, there for immediate purchase and wear lay these buttons, each emblazoned with a cute little piece of musical propaganda. And on one particular round, yolk-colored button, about the size of the bottom of a soda bottle, was printed the word "Whining," with a broad red line slashed diagonally across it. *No whining*, the button said. I bought four of these buttons, one for each member of the family.

These buttons referred specifically, I believe, to the necessity of practice, and the ever-present resistance to it, expressed usually in an elongated vocal plaint, which is preceded sometimes by an elongated audible moan. Often the vocal component is accompanied by a limp flap of the arms and a slight lifting of the eyebrows and head. Practice is, after all, concerted effort, and progress often slow and incremental; immediate gratifications rarely are available. Too often, especially early in the instruction of the music student, practice seems like a meaningless uncomfortable wandering in a wilderness, and the comforts and even joys available elsewhere seem all the more promising. How many Twinkles can one person play before he or she begins to cry? Complaints filled the air.

Or perhaps an inclination toward entropy keeps us all comfortably sedentary and resistant to effort. Movement requires overcoming this entropy, and movement begins slowly, uncertainly, and, often, creakily, and not without complaint. After all, what lies ahead is uncertain, and the effort to achieve it is almost always demanding. To make that effort involves energy and responsibility, and it is no small thing to accept these obligations and the efforts they entail. Even enthusiastic Suzuki people suggested that it was only at the age of about fifteen that the budding violinists would begin to take ownership of their own practices; before that time, they caution, the parent must assume responsibility for ensuring that every day that we fed the children, we also ensured that they practiced their instrument. As if not already sufficiently engaged, parents were encouraged to worry about the regularity of their children's practices, ordered to sit in on practices and lessons, even urged to take up an instrument to gain some facility and knowledge for the sake of the child's development. Our active involvement, we hoped, would minimize their complaint; we would satisfy their needs before they were spoken.[1] We would, as it were, sustain.

I wore that button every day during our camp stay, and for many days subsequently. I even hoped that its message would transcend the occasion of violin practice and enter into our family ethic. I imagined a world where whining was absent. And I recognized an interesting dynamic in complaint: when voiced, complaints did not simply refuse action, but in fact, enacted it. In inaction, the children in complaint became active. And while they complained, they could not practice. On one simple level, complaint as a form of action obviates the need for further and more concerted action. When I complain, I, too, express displeasure with a situation rather than effect change in it.[2] Certainly, there was more in the complaint than mere indolence; in the Freudian sense, the complaint was as overdetermined as any dream. When my children complained about practicing, they really worried about the effort they would need to make in practice; they worried that during the practice they might discover difficulty that required some greater effort to overcome; they worried that during practice they might have no questions to be asked and might become bored, even as they worried about the questions they would ask that required some response; they worried about what *else* they might be doing instead of practicing; and they worried that I would not care enough about them, and thus, allow them not to practice. That I would somehow in this enterprise abandon them, and in their activity they would not be sustained.

The presence of the complaint implies a complainer, and that complainer portrays himself or herself as somehow oppressed and a victim. Complaints posit a world that oppresses me and resists my desires; complaints posit a world more powerful than I am, and over which I have little control. Complaints maintain me in stasis: when I complain, I regret my present state rather than work to change it. I bemoan conditions, and engage in nostalgic reverie of a long-past Eden. I imagine with fondness returning to what never truly was. Complaining looks forward to a future that I am owed without my effort. Complaining protects me from moving into that indeterminate future: we prefer to stall in the known though unsatisfactory present. Complaining engages me in an inactive action that avoids substantive change. As an expression of displeasure, complaint calls for action, but need not necessarily lead to it. And complaints are all about us. If as Wordsworth says, "The world is too much with us late and soon…" it is complaint that keeps us so attached. If a complaint may be considered a voiced worry, then there is no dearth of objects for complaint. "A moment's thought," psychoanalyst Adam Phillips (1993, 55) notes, "will tells us what, if anything, we have to worry about tomorrow." I engage in complaint when I wish to hide the action I might take to obviate the complaint.

Perhaps, complaint is a holding action preparatory to action. Implicit in complaint is a potential for action along with a resistance to it. If there were no intent to act, complaint would be unnecessary. The complaint is a symptom of the need for support. When my children complained, it was because they resisted practice and not because they refused it. In the interim between complaint and act, I sustained them. That is why Shinichi Suzuki placed us in the practice rooms. That sustenance made action possible. When they practice, they did not require sustenance. Indeed, they practiced and ceased complaining.

Complaint is a resistance to responsibility. From the moment they left Egypt, the Israelites complained. With the Egyptian soldiers in close pursuit to recapture the newly freed slaves, the people accused God of leading them out of Egypt only to drown at the Red Sea. They were, as we know, saved when God parted the waters, though we understand that only Nachshon assumed responsibility for action.[3] Perhaps, miracles always

begin with the necessity for human action. Next, the Israelites complained
of a lack of fresh water; they were soon provided with a plenteous supply.
Wandering in the wilderness, not yet certain to where they were headed, the
newly freed Israelites grumbled about their lack of food, drink, and other
comforts. They complained bitterly that God had liberated them from
Egypt only to allow them to die in the desert. At least, they moaned, in
Egypt,[4] the food was plentiful and varied. There were cucumbers, onions,
maybe even chocolate cakes. Though in Egypt they were slaves, they cried,
at least there they would not starve. Better to be a slave in familiar territory,
they moaned, than to wander seemingly aimless in the wilderness.

Moses wonders aloud how long he must tolerate the murmurings of
these stiff-necked people. God tells him to be patient and to stop complaining; God tells Moses to assume responsibility and to handle things
himself. There seemed to be no end to their murmurings, and on at least
one occasion, in the incident with the Golden Calf, God, soul-weary of the
complainings, threatened to annihilate the entire lot and with Moses start
over again. Then Moses ascended to the top of Sinai to receive the words
of God, and, the people felt alone, seemingly leaderless and, I suspect, very
much afraid. They complained to Moses' brother, Aaron, decrying their
state, and demanded that Aaron build them "a god who shall go before us,
for that man Moses, who brought us from the land of Egypt—we do not
know what has happened to him" (*Exodus,* 32:1). Aaron builds them a
golden idol about which the people dance and celebrate, in express viola-
tion of the commandments which they had recently been given. God's
response to this idolatry is swift and brutal. God absolves God's self of
responsibility for the people, and in a threatened reverse of the creation,
God promises to cease sustaining the people and to obliterate them from
the earth. God says to Moses, "I see that this is a stiffnecked people. Now,
let Me be, that My anger may blaze forth against them and that I may
destroy them, and make of you a great nation" (32:10). God, too, has wea-
ried of sustaining these complainers. In a wonderful Midrash[5] on this
statement, it is said that in God's demand that Moses *let him be,* God
makes evident that if Moses would only make the effort, he might per-
suade God to recant his decree.[6] In a remarkable irony, while God complains, Moses sustains God. Indeed, through Moses' intercession God
renounces this purpose, pardons the people, and saves them from complete
destruction. Alas, the complaining did not cease, but the people continued
to be sustained in the wilderness. There was, it would seem, a greater purpose to their survival that all of their complaining could not efface; though
seemingly without end, their wandering in the wilderness was not without
an end. Perhaps it was for this learning that they continued to be sustained; their continued survival had purpose.

And so, during the wilderness wanderings, in response to the complaint of the hungry, thirsty, and frightened newly freed slaves, God daily brought the quail in with the prevailing winds, and caused the manna[7] to fall from Heaven. Thus, though wandering in the wilderness, the desperate Israelites found food and drink sufficient to sustain them in their puzzling journey. This manna, which it has been said was created on the eve of the very first Shabbat, was said also to vary for each in taste according to the particular need of each—to young men and women it tasted like bread, to the elderly it tasted like wafers made with honey, to sucklings it tasted like milk from their mothers' breast, to the sick it tasted like fine flour mingled with honey, and to the heathen it tasted as bitter as linseed. There, in the wilderness, God provided the people with daily sustenance; not only would they not perish, but as a result of the sustenance offered, they would thrive and learn that they had been freed for a purpose. They might have, though they did not, cease complaint.

But, perhaps, the manna did not come free of obligation: even free food required some payment. Moses commands the people that they must gather only that quantity of manna that could be eaten in a day. None must be kept overnight. And again, they complained: the Israelites "paid no attention to Moses; some of them left of it until morning, and it became infested with maggots and stank" (16:20). I suspect the camp smelt with rottenness and decay. I suspect the people complained of the stench they had themselves caused. Clearly, the manna provided would only sustain them *if* they followed the prescribed means for collecting it. Freedom, they would have to learn, was not empty time, but responsibility in time. If their lives had been once organized by having been slaves, the redeemed Israelites had now to learn the responsibilities of freedom. Freedom had to be learned, and during that learning, they required sustenance. During their wanderings in the desert they were sustained. I think they had to learn to cease complaining, and in their wanderings, learn to sustain themselves and others.

Sometimes, I think that the people's grumbling in the desert was an attempt to avoid assuming any responsibility for their lives. In the wilderness, their murmurings were expressions of displeasure and discontent; the people complained of their condition and sought from others relief. Whatever sustenance they received was considered their due, something even intrinsic to their being; they did not acknowledge a world outside of them. In this state, they could not learn, be creative, or be responsible. As might children expect, the wandering peoples would be continually cared for, but offer little care. As might children expect, the wandering peoples would be fed, but prefer not to help with the clean-up. With the Israelites' discomfort and responsibility came complaint. They had far to go and much to learn in the wilderness.

Complaint, as I have said, too often renders action unnecessary. God tells the Israelites that "I will make of you a nation of priests," but their grumbling reflects their resistance to this charge. They would seemingly prefer to continue to moan than to act, and to persist in a form of bondage. Even as my children, I suspect, have ambition to become violin virtuosos, they continue to complain about practice; and even we as teachers bemoan our condition and continue our practice. To do otherwise requires a great effort, and to that effort there develops great resistance. We do not feel sustained. William Pinar (2004, 33) writes, "Such efforts to reconstitute ourselves from merely school employees to private-and-public intellectuals cut against the grain not only of dominant school reform rhetoric dominated by right-wing intellectuals and politicians, they cut against the grain of traditional teacher education with its neglect of curriculum and its obsession with instruction and learning." We must begin to assert our authority in the face of adversity, and weather the storms we incite. We must not look too much without for support; we must ourselves practice what would sustain us in our wanderings. Bob Dylan writes, "I can tell you fancy, I can tell you plain, / You give something up for everything you gain / So since every pleasure has an edge of pain / Pay for your ticket and don't complain." The manna does not fall for free; if we are to be sustained, we must recognize and accept its cost.

Though we teachers continue to complain vigorously, we remain subjugated. We too often refuse to make payment. Ultimately, we complain because we feel powerless; we forget what it is we have to offer as payment. We complain because complaint is action without cost. We are not free and we remain enslaved. But I think we have to learn to pay for our tickets. Baruch Spinoza writes (1955, 187) that "Human infirmity in moderating and checking the emotions I name bondage: for, when a man is a prey to his emotions, he is not his own master, but lies at the mercy of fortune: so much so that he is often compelled, while seeing that which is better for him, to follow that which is worse." Such is the state of public education in the United States today: we know what we must do, but the way is long and the night is dark, and we are far from home. We wander in the wilderness, and we require some sustenance. We build our complaints about an imagined time when teachers held power and authority, when in fact, it is our work to construct such a time. Pinar (2004, 31) asserts that curriculum theorists might serve this function. He says "if we can teach, if we can make friends with some of those struggling in the schools, build bridges between the realms of theory and practice…if we create passages…to travel from here to there and back again, broadened, deepened, enlivened by the voyage, then we theorists might participate with subtlety and acumen in school reform." We, educators, theorists, and teachers together, like the Israelites in the desert, must learn to make payment for our freedoms, and cease to complain.

Every step into the unknown is a step into the wilderness. To engage in education, an enterprise in which I have spent my entire adult life, is always to walk into the unknown. There is, in this effort, place for fear, insecurity, resistance, and complaint. I have known complaint!! This is not the place for a philosophical disquisition on education; though, not surprisingly, some of that discussion may be found in this book. I have, after all, spent my life and much of my thought in schools. But certainly, I have come to appreciate that to truly learn one must first become truly lost and wander about in the necessary wilderness. We must accept the value of wildness and cease complaining about it. "Gardening," Thoreau (1834/1961, 61) says, "is civil and social, but it wants the vigor and freedom of the forest and the outlaw." Our schools have become too much cultivated garden and not enough wilderness; school culture does not bode well for that wildness that portends exploration, discovery, and freedom. In the schools our children no longer need be sustained. There may be, says Thoreau, "an excess of cultivation as well as of anything else, until civilization becomes pathetic" (61). Our schools have become nothing but civilized. I would advocate for the schools the presence of some wilderness, and if we would not sink into the culture of complaint, we must discover what would sustain us in our wanderings there.

We are, then, in our classrooms on errands into the wilderness. In his book *Errand into the Wilderness*, Perry Miller (1956/1964, 3) notes that there are at least two meanings of the word "errand." The first refers to a "a short journey on which an inferior is sent to convey a message or to perform a service for his superior." I might be sent on an errand by any number of persons: to a very large extent, this has been the organizing principle of teacher education programs in the United States. Teachers, very much considered inferiors, are directed to *go into that classroom and teach those children something*! What it is that teachers are to teach has been the subject of great debate since the beginnings of the common school in the beginning of the nineteenth century, but that teachers were sent on an errand remains incontestable. The established standards, set curricula, standardized tests, formulaic lesson plans, and innumerable courses concerning subject delivery and classroom management, attest to the accepted view of teachers as inferiors who are performing a service for superiors. I have engaged over my thirty-five years in great complaint of these directives, and have missed, I think now, too many opportunities to alter them. Complaint has become the only action teachers are left with, robbed as we are in this age of accountability, measurement, and government mandate of so much of our agency and our skills. We are not sustained, and we have no freedom to do our work; we are not sustained, and we cannot assume our responsibilities.

Eliminating wilderness has been the work of educational psychologists, who have mapped every thought and gesture of the child at each moment of his or her life; of curriculum developers, who have defined at what place each child must be at each moment of his or her schooling; and of too many professors of education who prepare preservice teachers by scripting every word and act they ought to speak. We have defined children's learning by adult scientific standards; the child comes into being in the discourse of the society of adults as a cognitive organization upon whom pedagogical practices must be enacted to ensure the education of the child. We have surveyed the field of education so thoroughly that one cannot step without tripping over a marker. We have abandoned all sense of awe and wonder for the illusion of certainties. Certainly, we have done away with wilderness. In our cultivating the wilderness, we have attempted to avoid expressions of dissatisfaction, though I think we have merely replaced one complaint with another. Writing in his journal, Thoreau says, "I have lately been surveying the Walden woods so extensively and minutely that I now see it mapped in my mind's eye…I fear this particular dry knowledge may affect my imagination and fancy, that it will not be easy to see so much wildness and native vigor there as formerly. No thicket will seem so unexplored now that I know that a stake and stones may be found in it" (January 1, 1858, 1240). Lacking wildness and in his actions lacking wilderness, the Walden woods become empty and hold little delight or interest. Thoreau says, "Dullness is only another name for tameness. It is the untamed, uncivilized, free, and wild thinking in *Hamlet*, in the *Iliad*, and in all the scriptures and mythologies that delights us,—not learned in the schools, not refined and polished by art" (November 15, 1850, I, 168). Without the wilderness, too much life becomes boring, a situation I will define as one in which no question might be asked. This type of errand is not productive. This errand leads not to freedom; here sustenance is absent. Our schools too often are dull, empty places offering little of interest or challenge.

The second definition of errand, Miller says, appeared toward the end of the Middle Ages, when errand came to mean the "actual business on which the actor goes, the purpose itself, the conscious intention in mind" (3). When Thoreau declares that he went to the woods to live deliberately, he announces his errand. That is, in this declaration, Thoreau proclaims that he was not sent to Walden on a mission by someone else, but rather, that his errand was to learn his mission. On his errand to front only the essential facts of life, Thoreau (1970, 222) would "see if I could not learn what it had to teach, and not, when I came to die, discover that I had not lived." Thoreau's errand is his purpose, and it is wholly an educational one. This errand means not to be sent, but to be found.

I think that it is this "errand" upon which the Israelites journeyed, and it is in this sense that I can say that in the classroom, I have been on an errand in the wilderness—I have acted to learn my purpose—and in that endeavor I have had no call to complain. I have been sustained. Oh, I will admit that too often I have ceased exploring and engaged in complaint, and I hope little of that complaint remains in this book, though all growth occurs through some conflict of which complaint is a symptom. With Thoreau, I have too often struggled with too much civilization in the schools, endeavoring there to find my voice and to help others discover theirs. Some of that struggle is the subject of this book.

In the course of some of those struggles, I came to discover that the voice I have been given to speak in and of the classrooms excluded me from the outset. Not only has the voice of the Jew been effectively silenced in the traditional discourses of Western culture, and especially that of education, but as Leslie Fiedler (1991, 16) notes, that traditional Western voice that has been taught as "our heritage was . . . theirs too, which is to say, Christian and therefore necessarily anti-Semitic." When I have spoken about curriculum in my classes, when I have practiced curriculum, I was constrained to enter the Greco-Roman and Christian conversation about curriculum theory and public schooling, and to employ the language of that realm. As a Jew, I spoke consistently from a position of silence. The socially recognizable yet socially illegitimate Jewish voice has been effectively silenced in the schools—except perhaps as stereotype, as in the teaching of Shakespeare's Shylock, or Dickens' Fagin, or as the Jew in Fitzgerald's *The Great Gatsby* or Hemingway's *The Sun Also Rises,* or the poetry of T. S. Eliot, Ezra Pound, and even Geoffrey Chaucer.

There are no theorists identified as Jews in the texts of curriculum studies, except perhaps in the single text, *Talmud, Curriculum and the Practical: Joseph Schwab and the Rabbis* (Block, 2004), and almost no reference is made in curriculum books to Jewish wisdom texts or philosophers. Judaism plays little part in the conceptualization of American education. For example, in J. Wesley Null's recent biography of Isaac Kandel, Null dismisses the influence of Judaism on Kandel's important work in curriculum by separating his work in Jewish education from his work in secular curriculum. Null (2007, 118) writes that "although his Judaism remained central to his life, Kandel did not continue to publish . . . on Jewish education," but focused "all his attention on comparative studies." Null seems to assume that the Judaism that was so central to Kandel's life, and from which his own philosophical foundations derived, had no influence on his mainstream work. Null discriminates between what he refers to as *rhetorical* scholarship, which is "moral in content," and therefore Jewish, and *descriptive* scholarship, which aspires to the presentation of facts and boasts a certain objectivity, and is therefore, not Jewish. Though Null acknowledges that Kandel engaged in both forms of scholarship, he does not situate Kandel's rhetorical scholarship in Jewish texts or ways of thought, and therefore, Kandel's scholarship seems essentially groundless and without context. Certainly, the work's foundation in Jewish thought remains unacknowledged. However, the sources of the descriptive scholarship as influenced by science, Taylorism, and scientific management, and the movements of accountability and efficiency, are well documented and eminently visible.

Indeed, there is in Null's biography even the strongest hint that Kandel's early failure to be appointed a regular faculty at Teachers College, though other friends and "fellow students of his" ascended to professor positions, resulted from institutional anti-Semitism, and that it was only after his work became more acceptable to Christian ears that Kandel received promotion.[8] Null writes, "Kilpatrick obviously was not accepting of Kandel's Jewish beliefs. If Kilpatrick's comments are any indication of what Kandel experienced generally during the thirteen years that he had to wait for a professorship at Teachers College, the fact that Kandel was Jewish had a significantly negative impact on his career" (102). Quoting George Bereday, who argues that Judaism was situated at the center of Kandel's work, Null concludes from this only that "Judaism must not have disappeared altogether" from Kandel's life, but he does not seem to be open to the idea that Kandel's Judaism undergirded his philosophical positions. Bereday (in Null, 2007, 271) says, "He was unmistakably a Jew. His Judaism was of the finest and of the staunchest. He was an unbending Rabbi, the mentor, the preacher . . . He was above all Jewish intellectually. He could not stand persecution and victimization of anybody anywhere. In this sense he was a staunch exponent of personal rights, an egalitarian. But he was also uncompromisingly intellectual and intolerant of anything less than the highest standards. In this sense he was an elitist." Nonetheless, when Null concludes his biography of Kandel, the influence of Kandel's Judaism on his work does not appear, nor when Null quotes "curriculum philosopher" Joseph Schwab, does he note that Schwab, too, was Jewish.

The Jewish voice has been forever absent from the field of curriculum studies. Sander Gilman (1991, 18) writes that for Western culture "The entire story of the Jews is reduced to a preamble to the coming of Christ," and silences the voice of the Jew, placing that story in an alien milieu. Christianity delegitimized, usurped, and reinscribed the Jewish voice, and altered forever pedagogical purpose and practice. The Apostle Paul's insistence on personal mediation (Jesus as mediator) replaced the textual mediation of Jewish practice. For Christianity, faith in Jesus substitutes for the immediacy of textual authority and the primacy of interpretation. Susan Handelman (1982, 84) writes "In place now of textual mediation is personal mediation through the figure of the new redeemer, whose presence has ended the long metonymy of Jewish history." For example, in *Romans* 10:4, Paul writes, "For Christ is the end of the law, that everyone who has faith may be justified." Here, interpretation ends with Jesus's word; the text is made flesh, and meaning is based in acceptance of Jesus's meaning. The whole pedagogical process that will become the Talmud and its methodology is denied.

In the Babylonian Talmud, *Temurah*, 16a, Rab Judah reported in the name of Samuel: "Three thousand traditional laws were forgotten during the period of mourning for Moses. They said to Joshua: 'Ask [through the holy Spirit] that the laws be recovered.' He [Joshua] replied: It is not in heaven." According to this tale, there is much work to be done here. The forgotten laws must be discovered here on earth: they are not in heaven. Indeed, Louis Finkelstein (in Handelman, 1982, 46) writes that Rabbinic thought holds that "...the texts is at once perfect and perpetually incomplete; that like the universe itself it was created to be a process rather than a system—a method of inquiry into the right, rather than a codified collection of answers." The following story, also in the Talmud speaks to this belief:

> Rab Judah said the name of Rab, When Moses ascended on high, he found the Holy One, blessed be He, engaged in affixing coronets to the letters. Said Moses, "Lord of the Universe, Who stays Thy hand?" He answered, "There will arise a man, at the end of many generations, Akiba b. Joseph by name, who will expound upon each tittle heaps and heaps of laws." "Lord of the Universe," Moses said, "permit me to see this man. God replied, "Turn around and see." Moses went and sat down behind eight rows [and listened to the discourses of law]. Not being able to follow their arguments he was ill at ease, but when they came to a certain subject and the disciples said to the master, "Whence do you know it?" and the latter replied, "It is a law given unto Moses at Sinai," Moses was comforted.

Though the law had been given to Moses he did not recognize its exegesis in Akiba's practice, but he is finally comforted in the identification of the text as the books that he had written down. No authority is authoritative; the process is all. Authority rests not in the text, but exists in the relationship between the text and its readers. Interpretation is central. Pedagogy concerns not what the text means, but how to create meaning with the text.

The Talmud, the text I will use throughout this book to think about issues of curriculum and education, is edited as a multilevel complicated conversation that took place over centuries, though that conversation appears in Talmud as a synchronous and seamless whole. In these arguments, the Rabbis do not necessarily attempt to reach a definite opinion about previous controversies or present issues, not even to definitively establish law and practice. Rather, it is their purpose to trouble meaning, and to open it to a variety of potential viewpoints. Indeed, the conversations in Talmud preserve and clarify the wealth and multiplicity of approaches to the problem at hand that have arisen over centuries and continents, even as Talmud develops and opens these issues to future meanings. Many sections of Talmud offer two or more opposing opinions without choosing between them. Moshe and Tova Hartman Halbertal (1998, 460) write, "Through refutation of challenges the Talmudic discussion advances multiple understandings of the Mishnah[9] rather than narrowing those interpretative options as far as possible. Within this dialectical process, initial opinions are inevitably reinterpreted, sometimes rephrased or articulated within a new context and therefore perceived in a novel light." It is not to arrive at a single authorized meaning that the Rabbis engage in study and conversation, but to maintain multiple interpretations, and to enable continued and open study of the text. In this way, there is in and for Talmud, no end to interpretation. Talmud offers a pedagogy of the question rather than an answer.

To speak as a Jew, then, is to argue against the Greco-Roman-Christian culture. To so read curriculum and education history against the grain would be to recover the hidden voices of Jewish traditions, beliefs, and practices in education; to read history against the grain would be to hear the Jewish voice in traditional pedagogical discourse and to uncover the suppression and translation of that voice; to read history against the grain would be, perhaps, to offer to American education an alternative pedagogy that might be truly transformative. This book, then, recounts *that* errand into the wilderness, even as the writing of it became my errand into the wilderness. I have in the classrooms and in the book sought out the wilderness and to realize my errand in it. I have even often thrived in it. Certainly, I was sustained.

And so the book is structured about my journeys into the wilderness—about the classroom, about teachers, and about my own concerns about immortality *as a teacher* and as a human being on this planet. This work is steeped in autobiography, and since I have been integrating Jewish wisdom texts into the discourse for upward of a dozen years, this manuscript is based in those texts and in my life.

I have also structured the book alluding to Prospero's announcement at the end of *The Tempest*, that upon his retirement to his beloved Milan, "every third thought will be my grave" (V, I, 311) Obviously, the final section of the book, on immortality, is most directly concerned with this issue, though I will raise this issue of the teachers' immortality in the very first chapter. I thought it interesting to organize the whole book around the aspects of my wanderings in the grade—the classroom; the brave—the teacher; and the grave—immortality. Thus, the book is divided into three sections: The first is titled "Every Third Thought Is of the Grade," and concerns the classroom and our work there. The second is "Every Third Thought Is of the Brave," and talks about teachers and their efforts, and the third section echoes Prospero, titled "Every Third Thought Is My Grave" with a slight structural change from the first two sections. This change is explained in the text. Of course, the whole structure is artificial because one cannot conceive the classroom without some people in it. In the same play, Miranda says, "Brave new world, that has such people in it," (V, I, 182–183) and this is my hope for the schools and the classrooms. Within each section there are chapters—four in the first section, two in the second section, and one in the third, all related, I hope, though somewhat loosely, to the theme of the section. Finally, as I have said, I have used Talmud as way to explore educational issues, not to offer sacred texts to answer fundamental questions, but as a means and model for opening the conversation to a greater complexity.

When I first conceptualized this book, I thought to call it *To Know My Song Well*. In this title I allude to Dylan's song, "A Hard Rain's Gonna Fall." There, to the question, "What'll you do now, my blue eyed son?" Dylan responds, "I'm going back out 'fore the rain starts a' falling / But I'll know my song well / Before I start singing." In this angry, prophetic song, Dylan describes a world on the edge of moral and physical apocalypse. Listing a litany of injustice, Dylan cries out "as if the sky were about to collapse" (Heschel, 1962, 4). Perhaps like those of the Hebrew prophets, Dylan's words were meant "to challenge and to defy and to cast out fear" (18), offering not prediction but hope. Heschel says that "The prophet is intent on intensifying responsibility, is impatient of excuse, contemptuous of pretense and self-pity. His tone, rarely sweet or caressing, is frequently consoling and disburdening: his words are often slashing, even horrid—designed to shock rather than to edify" (7). Having lived in this world and understood its wrongs, Dylan takes what he has experienced in this world, and "like a scream in the night," shocks the people to action. Very much like the biblical prophets, Dylan calls upon us to care for the widow, the orphan, and the stranger in our midst.

I think that the song that Dylan would sing is the one that he has learned all too well from his wanderings in the world. Dylan sings of heading out into the wilderness, not, like Huck Finn, ahead of "Sivilization," but into the wilderness that is "Sivilization." For Dylan, there does not seem to be any inclination that it is complaint of which he will sing. Indeed, to know one's song well derives from a knowledge, a surety, and a power. "Not the fact of his having been affected, but the fact of his having received a power to affect others is supreme in [the prophet's] existence. His sense of election and personal endowment is overshadowed by his sense of a history-shaping power" (Heschel, 1962, 21). Alas, we teachers are not prophets. Another definition of wilderness, dating from 1588, refers to wilderness as "a mingled, confused, or vast Assemblage of persons or things." I think this might be an apt description of the place I have spent my life in. In this book, I have tried to describe the pleasures and the pain, but I am hoping it is free from complaint, and that it offers some proof of payment.

But there is an ambiguity to my title. In one sense, I express in it a sense of completion: I express in this book what it means to know my song well. Having completed my sixth decade, and having spent thirty-five years in the classroom, I wanted to think that I knew my song well, and that I had something about which to sing. But in another sense, *to know my song well* suggests some pedagogical purpose in these pages: the writing has helped me know my song well. I have been sustained.

In Torah, immediately following the narrative of the provision of manna, we read this command:

> Moses said, This is what the Lord has commanded: Let one *omer* of it be kept throughout the ages, in order that they may see the bread that I fed you in the wilderness when I brought you out from the land of Egypt. And Moses said to Aaron, "Take a jar, put one *omer* of manna in it, and place it before the Lord, to be kept throughout the ages." As the Lord had commanded Moses, Aaron placed it before the Pact, to be kept.

The manna that could not last beyond morning was meant to last forever! This measure of manna in the jar holds meaning: See how you were maintained in the paths toward your freedom. Every step into the wilderness is a step toward freedom; freedom occurs from the wanderings in the wilderness. But today, when I search for this jar, it is not to be found. It no longer exists. This is a ritual no longer practiced. I think that there is something important lost in that loss, and I would recover that jar now. In this time of my plenty, I would know by what I was sustained in the wilderness. In this time of my plenty, that jar might teach me how to sustain others. I think I would now make use of that jar in the classroom. Into that jar I would place an *omer* of learning to be kept throughout the ages.

This book is my jar, and the writing the manna that has sustained me. I take it out and give it to you.

Part 1

Every Third Thought Is of the Grade

Chapter 1

Why Be a Teacher

I am finishing my sixth decade. Every third thought is of the grade.

Almost every day in each of the past thirty-five years I have daily traveled to the school in my role as teacher, and for the remainder of those years I attended as a student. Becoming a student was simple: my parents sent me to school and there they let me in. Apparently, I have never found my way out! In school as a student, I achieved some success. That is, I never received suspension, never went to the principal's office, never was beaten up, and failed not a single course. I am certain that I learned something there, but what it exactly was I cannot say except in the vague generic sense of reading, writing, and arithmetic. Mostly what I must have learned in school was that I somehow liked being there, and that I liked learning, and though I suffered through not a few Mr. Hyde's, and Mr. Jones's, and Ms. Bueschel's, I also celebrated learning with Mr. Matienzo (whom I loved), and Mr. Meissner (whom I affectionately called "Peeps"), and Mrs. Seaman. The last was my fourth-grade teacher and I remember thinking her last name amusing, considering how odd it was that a gray-haired female teacher should have a name meant for a young male sailor. Then, neither I nor anyone else with whom I was familiar knew about male ejaculate, and so the lewd reference we might have ascribed to her did not occur in *my* social circle. Indeed, I was relatively socially inept, and except for a short fling in fourth grade with Marjorie Eisenberg (to whom I had given a recently gifted ID bracelet) and which relationship lasted twenty-four nerve-wracking hours, I did not have any girl or boy friends. I did, however, have boy and girl *friends*, and never experienced isolation or existential loneliness in school, though in high school I dressed in black turtlenecks and read existentialist literature, and loved when Renee Lerner referred to me as her "poet." Though I was never a stellar student, and do not now remember struggling very much academically (indeed, I can't remember studying at all!), I earned all A's and B's and mostly O's and S's on my report cards. In school, I learned. I learned, in school. It was a beginning.

When (and where) I went to high school, many in my social set desired to be doctors and lawyers. I recall that one in our circle chose Annapolis and a military stint for his postsecondary experience; back then, though we were paying some attention, we were not overly concerned. The Cuban missile crisis had ended without the world's end, and the cancer of the Vietnam War had not yet metastasized. Graduate degrees were exclusively of the MD (which included dentistry, but not vetinariancy) or LL.D. variety. In fact, I don't recall being aware of the existence of either master's or doctoral degrees; that academic direction did not exist for me. Indeed, I cannot certainly say that I ever connected academia with a profession—I enjoyed learning, but learning per se seemed hardly possible as a career choice. No one then spoke of the engagement in scholarly pursuits, and none of us entertained teaching as a career. Indeed, only a few girls, and none of the cheerleaders, belonged to the Future Teachers of American Club. That my teachers had college degrees, and that some might even have earned advanced graduate degrees did not cross my mind. It was the past *nonacademic* lives of some of them that had for us meaning and gave to these teachers substance: we knew that Mr. Tobin had been a policeman, and that Mr. Harrington had been in the FBI! We speculated when Miss Poltorak would ever get married (who would marry her, she was so mean!), and we noted that Mrs. Rieff could often be seen food shopping *for her family* at the local Waldbaums. I'm not sure that I ever connected teaching with academia, or of teaching with intellectual engagement or achievement. It just never crossed my mind that I would ever choose to be a teacher, though looking back now, with not a little irony, I see that during that time, all of my favorite adults were teachers and baseball players. But I think now that this was part of the beginnings of my choice. I didn't know any baseball players, but remained familiar with the teachers. They signed my yearbook, and promised me success in learning and even from learning, though none mentioned success *in* academia.

Like many of my friends, I planned to be a medical doctor. I went to college to realize this goal—what else was college for? If medicine was a *dream*, I don't think it was *my* dream; rather, it seemed the best of the available few choices my culture offered. Though I adored the heroism of the doctors popularized in the media, I personally couldn't stand the sight of blood, couldn't remove splinters, and most of all, was not particularly suited for the rigor of its study. I dreaded math and science and laboratories. It was the ideal doctor I aspired to be, whose white robe never got sullied, and whose efforts always succeeded with almost no travail.

I wanted to become Dr. Schweitzer without having to travel to Africa, to be Dr. Casey in the operating room without getting sprayed with blood, and to suffer like Arrowsmith struggling in the laboratory while his adoring wife waited up all night for her hero's return. My desire, as with so many of my other good desires, had nothing to do with reality, and I had neither the facility nor the drive to transform this desire to an accomplishment. Enrolled in the pre-med program at Roanoke College, I miserably failed the first math test that was purportedly a review of what I had supposedly learned in high school, but that I obviously had not yet sufficiently mastered. If they'd asked properly, I could have told them that. I dropped back a level to more basic mathematics and missed nary a class. During the final exam, I sat for three hours with nothing to do, and received a D in the class for effort. Dr. Walpole admired my persistence, but he despaired, I think, of my ability.

I persisted relentlessly, however, but I was not doing all that much better in my science classes. Though I had adored Mr. Meissner (Peeps), my high school biology teacher, and had dissected earth worms, and fetal pigs and sand sharks immediately before lunch period, I had little affinity for the subject, really. I was not drawn to it, and asked few interesting questions. Perhaps I was not inspired to do so, but I do not recall having any great desire then for this knowledge. Of high school chemistry all I can recall was not balancing equations and Mr. Medoff's odd behaviors, and the only test I passed all year in Mr. Martin's physics class was the end-term State Regent's exam. I scored a 76 when the passing grade was 70. Those early courses were merely curricular tracks on which I had to run, and the early college science courses impressed me more as a means to an end, and I was not interested in means. I only knew about ends. In college, the chemistry incrementally became more complex and decidedly more mathematical: there was qualitative analysis, and Dr. Bondurant's quantitative analysis. Dr. Bondurant was a rotund man who smelled of sweat and of the huge cigars he smoked incessantly; always out of breath, he lumbered about the laboratory and the calculator room in his wrinkled blue suit, jacket flung open on his stained white shirt, his tie either flying loose or caught on his left shoulder, shaking his head despairingly behind my computations. Of course, then there were no computers; the slide rule was still the primary tool of calculation, though I wouldn't be caught dead with one hanging from my belt. Besides, I had never in high school learned how to use the slide rule.

My college advisor suggested I next register for physics, organic chemistry, and calculus for the fall semester, and one day in the spring, I was sitting reading Wordsworth under a tree and I switched majors from pre-med to English. I didn't exactly understand "The Prelude," but I didn't understand it more aesthetically and more contentedly than everything I didn't understand occurring in the science wing. I preferred reading, anyway, and I was finding great success in my literature courses. And besides, there were very many more girls in the liberal arts. I became an English major. It was as easy as changing shirts; I think then I believed that it was all classroom anyway. What I was going to do with this degree in English, I gave not a thought. All of my ends had turned to means. Clearly defined objectives have never been my forte. This must have been something of a beginning.

I am thinking that there is a difference between beginning and origin. There is the beginning of my teaching career, and there is its origin. In the beginning, there was Jericho High School where I learned, I think, and what I learned mostly was the comfort and warmth of learning, even though I failed at learning, as say, in most of my science and mathematics classes. School was a safe place, and in those beginnings school was where I found not necessarily social ease, for adolescence and puberty cannot be avoided in school, but an intellectual security that nurtured me. As I finish my sixth decade, I return to these beginnings and observed not a confident social boy, but a somewhat insecure and frightened individual, seriously unaware how I was physically maintained, but certain that I felt emotionally abandoned at home and somewhat alone in the social world, ignorant of most practical knowledge, and unaware how to kiss a girl. In this beginning, it was in literature that I found support for my own feelings, and in school, and mostly among the teachers, I found some community. In the classroom and in learning I felt nurtured. In the womb of the classroom, in the beginning, I felt *always* safe. I have always loved being in school.

In the beginning, also, there was the war, and America's resident and prospective intelligentsia (!) were offered exemption from service in the armed forces if they remained somehow in school, either as undergraduate student (alas, I was graduating as an English major!) or as teacher, the latter considered a service of vital national interest. In the summer of 1968, with graduation just one year away, I enrolled in classes in education at Hofstra University in hopes of earning licensure from the state of New York and draft exemption from the U.S. government. During that summer, in the morning I studied education, and in the afternoons I drove to Jones' Beach. I recall the standard course fare, though little of its substance: that summer I took Foundations of Education, Teaching Reading in the Content Area, and Educational Psychology. Of these, I recall absolutely nothing but their names, and I have not retained a single note or text from the classes. I think there were one or two other courses of which even the name escapes me, but there was one, to which I shall soon return, to which I attribute the origin of my life as a teacher. But when I returned to

Roanoke College in the fall, I enrolled in yet another innocuous and now nameless education course, and during that spring semester of my senior year, I student-taught English at a high school which prefers to remain anonymous. The only extant evidence that I had any influence whatsoever on anyone came when one of the senior girls asked me to be her Prom date. I respectfully declined.

In the summer of 1969, following graduation and prior to the discovery by my draft board of my driftless and nonacademic state, I called schools on Long Island until I found one that had an opening for an English teacher. The job paid $7,300. They offered me the position, and I took it. I was again draft exempt. On September 2, 1969, dressed suitably in a sport coat and tie, I walked into my first class at Sayville High School, in Sayville New York. To each student, I handed out a literature anthology and a Warriner's grammar book. I stood before the class, and I began to sweat profusely. It was a beginning.

I did not last long at Sayville High School, but I have stayed in the teaching profession. I do not intend to romantically situate the origin of my engagement with teaching somewhere in my childhood and adolescent love of school. That would be all fiction. And though learning has always interested me, indeed, interests me viscerally yet, as I have said, learning didn't send me into the classroom as a teacher, nor did it originally maintain me there. It is more accurate, however, to associate my origin in the profession with my enrollment in Franklin Stein's methods class, Teaching Secondary School English. This was the final course in the sequence I took during the glorious, sunburned summer of 1968. By the time I crossed the lintels of his doorway I was very tan, and had met not a few girls during my afternoons at the beach. At the front of the room stood a man, thin, darkly bearded, smiling, and very animated. On his reading list was Marshall McLuhan's *Understanding Media*, Stephen Dunning's edited poetry anthology, *Reflections on a Gift of Watermelon Pickle*, and several other literary texts whose titles now escape me. I grew concerned: where were the methods? If I was going to become a teacher, I had to acquire an algorithm! I did see on the syllabus a tome entitled, *How to Teach Secondary School English*, or something like that, but on that first day, Stein said we needn't buy that text if we weren't so inclined, though it was not a bad recipe/sourcebook for a rainy Monday morning. I purchased it immediately, but I did not open it until the first Monday rainy morning in Sayville, and discovered Stein was correct. The recipes were, indeed, satisfactory, though I always seemed to be lacking one of the essential ingredients.

Truthfully, I don't recall at this remove Stein's complete reading list, but I do recall the unorthodoxy of it, and the books I have named were certainly on the syllabus. It was not that the books were uninteresting—I found them not unlike the texts I had always been assigned as a literature major, but they offered me perspectives absent from my undergraduate curriculum. And the reading list was not all that was not kosher about that education class. Every day Stein produced another experience that I always enjoyed but didn't understand. One day we watched the film *Dream of Wild Horses*. We *talked* about the film, but not what to do with it in the classroom. One day Stein showed us slides, but he insisted on keeping them out of focus. We wrote down what we thought we saw, and shared our experiences. We did not consider the classroom. One day Stein asked us to put our heads down on our desks with our eyes closed. He put a record on the turntable: Tchaikovsky's "Francesca di Rimini." He would stop it occasionally and read from *The Diary of Anne Frank*. Sometimes he would tap one of us on the shoulder and ask us to read from the *Diary*. We talked about how the literature gave a structure to the musical work, and how the music gave a new depth to the *Diary*. I actually saw some potential connection between this event and the classroom, though later I would discover that following this recipe, I yet lacked an essential ingredient.

Stein never told us to "do this," or "do that." He referred to every class as "a happening." This was a 1960s term that referred to a communal event categorized by spontaneity, energy, creativity, and often, drugs. A "happening" was a self-contained event, and did not require history or continuity. A happening "happened," but could not be planned. During the previous summer session, I had learned that continuity was essential to the classroom and its management, and that carefully organized and sequenced plans were necessary to ensure that a proper education would occur. I learned to write tightly scripted, developmental lessons plans, and to organize whole weeks and months with meaningful, educative activities. Franklin Stein's practice contradicted vigorously everything I had learned. I grew very confused, but not unhappy: I couldn't wait for each class to begin to see what would happen! At the final exam, a disheveled and decidedly unsunburned young man pasted broken pretzel pieces into his blue book. I think he earned an A. I filled two blue books with a carefully planned, month-long, multidisciplinary integrated unit on *The Diary of Anne Frank*. I also earned an A. I have never used it—the plan or the A.

If I had learned the comfort and warmth of study in my high school classrooms, and if I gained emotional legitimacy and support in my engagement in literature, and if my burgeoning expertise in analysis helped me learn to shape and to understand my experiences, Franklin Stein taught me the excitement and wonder of learning. Today I know that to stand in the classroom in study is to stand in awe and wonder; to stand in the classroom in this manner is a form of prayer. Franklin Stein taught me that one could pray in the classroom. And he taught me how to pray there. Of course, when I walked into my first class at Sayville High School I was constrained to do nothing but practice the learning from the first summer session: I lectured, I asked scripted questions, I gave quizzes and tests and sat students in rows. I sometimes sent a student or three to the principal's office. I taught high school English. Whatever life I experienced then occurred outside of the classroom; sometimes there were even students with me.

But I had experienced something in Stein's class that I would not ignore at the risk of losing the classroom. It was not in 1968 nor 1969 that I understood Stein's pedagogy. It was in the mid-1970s, when I was working at Great Neck North High School, on Long Island, and teaching Thoreau's *Walden* sitting atop a four-drawer file cabinet, wearing a hat labeled "Thoreau," with students sitting irregularly in the classroom, along the wall, slumped in their chairs or stretched out on the floor or the window sill, and the principal, dear Mr. Bernard Ludwig, sitting alone at a desk in the middle of the classroom during his scheduled observation of my classroom, that I understood what Stein had been doing: he was teaching us not how to teach, but rather, he was teaching us as we ought to teach. We had made a "happening," and we were all very happy. And this joy was the origin of my teaching career. I have sought ever since to be worthy of Stein's pedagogy.

Professor Elliot Eisner (2006, 44) announces in a recent article in *Educational Leadership* that, "Each year, thousands of new teachers enter the field. Almost all seek deep satisfaction from the processes of teaching. Among the many satisfactions that exist, I would like to describe six." I appreciate his summary, though I have my doubts as to its relevance. His article offers a standard, idealized menu of some positive aspects of the teaching profession, but does not address the daily realities teachers in the field regularly confront. That is, to my mind, Eisner's list appeals to the rewards of teaching reflected upon in moments of tranquility after a lifetime in the classroom, but suggests after a lifetime in the classroom[1] none of the profession's difficulties daily experienced there. Eisner's list might lead the reader to believe that if the teacher does not experience the same satisfactions in the classroom as he, then s/he has only herself to blame. "How we teach," Eisner writes, "is related to achieving the deep satisfactions of teaching" (46). In the schools toward which Eisner's list points—schools of quality and higher socioeconomic status, schools like Stanford University—the good teacher would have to be supremely incompetent not to be satisfied.[2] But Eisner's list actually offers me no more information about why we teach than when I ask my doctor why he studied medicine to become a physician, and he answers, "To help people!"

According to Eisner, teaching allows him to participate in the world of great ideas, to realize a form of immortality, and to enact performance, or as Eisner describes it, "to play your own cello" (44). Teaching, Eisner continues, provides opportunities to create and participate in forms of aesthetic experience, to experience and represent a passion for learning, and finally, to make a difference in students' lives.[3] These are all admirable explanations for a commitment to the teaching profession, and I have every reason to believe that Eisner participates passionately in the profession for each of the reasons he enthusiastically offers. Of course, it strikes me now (as it does, actually, at other times as well) that the ends of teaching ought to enable *everyone* to experience in his or her own life *regardless of profession* the joys Eisner ascribes to teaching. Not that everyone must become a teacher, mind you; rather, I consider that what we actually do in the classroom is to *enact* our daily joys that include, but are not limited to, the six satisfactions of teaching that Eisner names, in the hopes that others will come to experience in their own lives these satisfactions that Eisner attributes to teaching. Perhaps this goal ought to serve for every classroom as the rationale for our content which must be not, therefore, vocation-specific. John Dewey (1964, 248) writes,

> Ultimately, the question of values and a standard of values is the moral question of the organization of the interests of life. Educationally, the question concerns that organization of schools, materials, and methods which will operate to achieve breadth and richness of experience. How shall we secure breadth of outlook without sacrificing efficiency of execution? How shall we secure the diversity of interests, without paying the price of isolation? How shall the individual be rendered executive *in* his intelligence instead of at the cost of his intelligence. How shall art, science, and politics reenforce one another in an enriched temper of mind instead of constituting ends pursued at one another's expense? How can the interests of life and the studies which enforce them enrich the common experience of men instead of dividing men from one another.   (Italics in original)

Were education to be organized around these criteria, then everyone, I think, should realize in his or her profession (and life!) satisfactions similar to those Eisner derives from teaching. Ironically, then, it seems that it is ourselves *as teachers* that we offer *as curriculum*, though in the current political climate, it is not safe to acknowledge (or be!) this in public.[4] Indeed, many of us even in private suppress such awarenesses. And besides, how will we test this?

Nevertheless, I acknowledge that I do not know anywhere another summary list attempting to define a commitment to teaching *and* learning *and* living. That is, may we all derive from *all* our lives the satisfactions that

Eisner derives from teaching. May we all teach so that all may derive from their lives the satisfactions that Eisner derives from teaching. Certainly, Eisner's account defines the rationale by which I have explained the thirty-five years of my teaching career in the public schools and public universities in the United States. And now, as the sun sets on my classrooms,[5] I am reexamining for myself each of Eisner's explanations. Rather, I am experiencing a failure to understand exactly what Eisner might have meant—what I over the years might have meant—by each "satisfaction" of teaching, and I now want to understand what I have been doing for the past thirty-five years. And I desire in this sunset to experience, perhaps, a purer and brighter light than I saw when first the sun rose. For me, now, the items on Eisner's list represent an ex post facto explanation derived from a lifetime in the classroom, and do not necessarily serve to encourage any person to enter the profession except in some idealized fashion. And when I query my students about their motives for becoming a teacher, the only one they voice that approaches an item on Eisner's list concerns the last one: making a difference in student's lives. However, not a single student can define with much precision what that desire might mean.

Indeed, when I probe their response just a bit further, I realize that my students' desire to enter the profession is often in the service of repairing the damage done to them by teachers. I am alarmed to consider that I am yet teacher to them, and I am presently alarmed that my satisfactions in the classroom might be achieved in their silence, and even, their oppression. And it strikes me, too, that in their desire to be teachers, they want to be that for which they have no model, and they must if they are to succeed as teachers inevitably strike out on their own *without my assistance.* Thoreau (2001, 226) writes, "If you are ready to leave father and mother, and brother and sister, and wife and child and friends, and never see them again,—if you have paid your debts, and made your will, and settled all your affairs, and are a free man, then you are ready for a walk." I think I would have my students take that walk, but in their classes, these students are learning to follow straight-cut paths, state mandates, guidelines and practices, and are forever denied their own practice. I think, *contra* Eisner, that we are training our teachers to be faithful slaves. "Academies," Spinoza (2005, 369) writes, "that are founded at the public expense, are instituted not so much to cultivate men's natural abilities as to restrain them." By the measure of our schools, we are not so free, and the satisfactions that Eisner describes will not be available to future teachers. My students are not free enough to be saunterers, nor do we teach that freedom.

Or sometimes my students express in their desire to be teachers the wish to be just like their teacher in elementary school or high school, in which case, they desire to become someone else, and will not learn to be themselves in the classroom. That way madness lies; the pedagogy here is antithetical to the realization of any of Eisner's satisfactions. Thoreau writes in his journal (December 31, 1859):

> How vain it is to teach youth, or anybody, truths! They can learn them after their own fashion, and when they are ready. I do not mean by this to condemn our system of education, but to show what it amounts to. A hundred boys at college are drilled in physics and metaphysics, languages, etc. There *may* be one or two in each hundred, prematurely old perchance, who approaches the subject from a similar point of view to his teachers, but as for the rest, and most promising, it is like agricultural chemistry to so many Indians. They get a valuable drilling, it *may* be, but they do not learn what you profess to teach. They at most only learn where the arsenal is, in case they should ever want to use any of its weapons.   (Italics in original)

My students seek my truths rather than their own. They would be teachers who have not learned how to learn. I sometimes point this paradox out to my students, but they do not attend to me. They want to know what they have to do to get an "A" in the course.

Ironically, in the very same issue of *Educational Leadership*, Scott Mandel (2006, 66) writes that too many teachers are leaving the profession after their first year in the classroom because of the stresses they suffer there. If, Elliot Eisner (2006, 44) announces in his piece in *Educational Leadership* that, "Each year, thousands of new teachers enter the field," Scott Mandel charges that one-third of all first-year teachers leave the profession at the end of the school term. "New teachers," Mandel writes, "have one basic goal in mind—survival." As my grandmother might say, "Hoo, hah!" These first-year teachers will never have the opportunity to realize the satisfactions Eisner lists as the rewards of the teaching profession—great ideas, immortality, performance, artistry, a passion for learning, and making a difference. Indeed, I suspect that these first-year teachers cannot even anticipate *ever* realizing these pleasures, and that is why they exit the profession.

Of course, even seasoned teachers often name survival as their primary desire, and though they maintain fantasies of realizing the satisfactions Eisner heralds for the classroom, too often they realize only frustration and pain.[6] In 1999, I returned to the middle school classroom. In the fall of that year, I taught Social Studies—American History, to be precise—in a private school in St. Paul, Minnesota. It was the most difficult work I had done in years. For almost two decades I had been an English teacher (perhaps experiencing the rewards Eisner names, but not consciously thinking of my rewards), and for the next decade, I had immersed myself in curriculum theorizing; I wrote voluminously and extensively about the desperate need to rethink educational practices in our schools. In the present instance, I wanted very much to help these Middle School students begin to think like an historian, a term I defined as "someone who always knows that there is another story yet to be told." I wanted to use the discourses of popular culture—folk song and original source material—to give expression to the silenced voices in American history and to interrogate some of the stories to which students already gave credence. I wanted learning to be challenging, fun, and, well, empowering. I mean, to use the words of Henry Giroux (1992), I wanted to "resurrect traditions and social memories that provide a new way of reading history and reclaiming power and identity...creating new languages and social practices that connect rather than separate education and cultural work from every day life." I mean, I wanted to change the world.

And how should I presume?

I miserably failed; I failed miserably. In one class studied Rashid, a highly motivated sensitive young man who actually said, "I enjoy studying with you," and in that same class, studied also his sister, Phyllis, who could joyously miss days of school on end, and then not understand why she was unaware what topic we were at present discussing. Rashid completed two books before his sister picked up her first. In one class sat William, an intelligent and involved student who suffered from epilepsy and was scheduled at the Mayo clinic for delicate brain surgery to relieve his symptoms; there was also Rochelle whose mother has been recently institutionalized and whose younger brother's behavior was so extreme that the family was having difficulty placing him in any school at all. How shall I consider these students' homeworks turned in late and to my mind, carelessly prepared? In one class sat Martha Will about whom I wrote, "Martha is highly distractable. The demands of her social life seem to be a priority," and in that same class sat also Sophia Ginzburg, about whom I wrote, "Every teacher deserves a student such as Sophie. Her powerful intellect and imagination stimulate and challenge." How shall I speak to both at once? Or even separately? In one class was Richard Overby about whom I wrote, "Richard has a very difficult time in class. He does not seem to have a sense of inner control and is disruptive and highly distracted;" and Jamie Key, about whom I wrote, "Jamie is one of the most delightful students I have known. Though quiet, she is an enthusiastic learner...Her writing displays a sophistication special to middle school experience." Is it American history that would benefit Richard now or a sense of inner control? Is it a lesson in history or a warm heart that might comfort Rochelle? How do I begin to address the concerns of both children—and that of the fifteen other children in the class—and satisfy my own beliefs and practices concerning the American history classroom I imagined.

I sit here now recollecting emotion in moments of tranquility and wonder incredulously how I could have ever believed that I might have succeeded! Or that my success would have been so evident that even I would have been pleased? And if I *was* pleased, perhaps it was me of whom I was originally thinking: violence, I recall, is to act as if I were alone to act!! How could I expect that all of these students should all feel the same about school at the same moment as that moment of the class in which I was teaching? And that somehow these students would be intrigued to exactly the same degree and coincident with the present material as I had offered. How could I expect that all of these students would be willing to learn with me or from me? What was Hecuba to them or they to Hecuba that they would weep for her? What right did I have to demand very much intellectual struggle from them when for many just getting to school was an achievement far in excess of my seventy-mile drive! How could I have expected all of these students to be like me in my idealism and occupy my idealistic image? Transform society? Ha!

Each and every day I was overwhelmed by the insurmountable effects of society upon these children. These were mostly privileged children whose parents could afford tuition to a private school, but their situations were obviously influenced by the personal and family psychological situations and conditions I could not even attempt to make sense of, much less begin to relieve. Tolstoy had said that all happy families were alike, but that every unhappy family is unique. He was wrong: every type of happiness, too, is different. Weren't Rashid and his sister both apparently happy and yet the product of the same home? Or shall I abandon all belief in authentic happiness and assume it to be yet another social mask? There was something intrinsically wrong with the whole notion of the classroom as it architecturally, structurally, socially, intellectually, and academically existed. How could I possibly deal with all of these personalities and still engage in American history learning? There were days when all I wanted to do was to teach these children some sense of inner control. There were days when all I wanted was to get out of there alive. Oh, do not mistake me, I wanted them to learn some American history,[7] I wanted their stories to assume rightful places in their lives and add to the discourses of American history; I wanted their stories to intervene in the monolithic American legend. I wanted to be a great teacher. How could I presume? How many eyes must one man have before he can see people cry? I have rarely felt as isolated in a classroom as I did during those months. What influence I might have had very little to do with what I taught. In that classroom I experienced at various times each of the joys Eisner lists as motives for teaching, but, too, too often it was survival I sought. And the road was so long and hard. I was not even a first-year teacher.

Henry Giroux calls teachers cultural workers; as such, they must discover the local custom because for the most part, as James McMurty sings, "I'm not from here, I just live here." Giroux (1992) valorizes this aspect of the pedagogical encounter as border crossings. And how should I presume? I wonder, what serves as my passport? Aren't border crossings also interpretable as trespassing? Woody Guthrie laments that those who cross may be known by no names, except "deportees."

I think the classroom is a very difficult place in which to reside, and the physical, intellectual, and psychological demands it makes on the teacher are wholly underestimated, if not completely ignored. Certainly, they are not evident in Eisner's paean to the profession. His classes, to a large extent, are all elective, and at the graduate level: I do not have a sense what his response might be if one (or none!) in his class was interested in the great ideas. The tensions that arise between the discrepancy of what I say in the classroom and what I do in it, between what I intend to say and what is actually said in the classroom, the conflict between teaching to live and living to teach, the split between content and context, trouble my rest and disturb my practice. Rabbi Ishmael (*Pirke Avot*, 4:6) said that "One who learns in order to teach is afforded the opportunity to learn and to teach. Whereas one who learns in order to practice is afforded the opportunity to learn, to teach, to observe, and to practice." But what that practice means with regard to teaching has become today quite problematic: the state mandates the script, and the textbooks the texts. To practice becomes recitation, and to learn merely imitation. I teach now in a school that educates teachers: the students there demand that I teach them "how to teach." Tell me what to do, they plead, and we will do it. "Just please, give us direction." I wonder if I am educating my students to teach or to practice. If the former, I teach a craft, but if the latter, I offer a vision, an art, a way of life. I think there are no set rules for such practice that I could transmit, and I seem doomed to a palpable failure that belies Eisner's satisfactions. Rabbi Ishmael's statement suggests a qualitative difference between learning to teach and learning to practice, and Dewey teaches me that this same dualism must be exploded if I am to proceed in teaching, practicing, observing, and learning. I would stand with Dewey and Rabbi Ishmael. To bring to an end this dualism is to stand confident in the world and in the classroom. I think that what I have always sought in the classroom is ethics and not pedagogy, but that word does not appear on Eisner's list.

Eisner's listing does not account for my experience in that middle school class, nor the ethics I seek. Eisner's list does not speak to the experience of those first-year teachers who leave after their initial year in the classroom, nor to the 50 percent of teachers who exit the profession within the first five years of their careers. Eisner's list does not address basic issues of survival. I wonder how one might teach survival, and by what measure these skills might be assessed. Perhaps survival is *already* a form of art and immortality.

In his essay "Walking," Thoreau (2001, 225) announces: "I want to make an extreme statement, if so I may make an emphatic one, for there are enough champions of civilization: the minister, the school-committee, and every one of you will take care of that." The rhetoric of the school-committees—national, state and local—have much to say for civilization, or at least, for their version of it. They have over the years defined what knowledge is of most worth and mandated its transmission. In the service to the state, teachers have been transformed into marionettes at the least, and slaves at the worst: it is not inopportune to recall here that an original meaning of "pedagogue" referred to the slave that walked the child to the school.

The satisfactions that Eisner attributes to the teaching profession are not available to the teacher because the teacher does not freely practice in the classroom. These committees have transformed the play by demanding that our students "study a speech of some dozen or sixteen lines which I would set down and insert in't" (II.ii.525–527). This speech serves purposes that the players could never know—indeed, must not know—and over which they have no control. In Shakespeare's *Hamlet*, when the players perform the edited drama *The Murder of Gonzago* before King Claudius, they are rightly bewildered, even angered by the audience response to their performance: "Hey, I said what I was told to say! What happened?" So, too, in our classrooms: from the often puzzled, often angry responses of students teachers must wonder, *What did I say?* Teachers are turned into hired servants. Condemned to speak inadequate ideas, they understand neither the source nor logic of the ideas they deliver, nor do they know those ideas as stemming from their own nature. Teachers are not considered sources of knowledge, but mere transporters of it. They are transformed into teamsters for the school committees, the minister, and other representatives of "civilization." Nor are teachers able to teach to students the capacity to develop adequate ideas the teachers are not permitted themselves to develop. Teachers themselves wonder, "What did I say?" As Spinoza (1955, 40) argued, in *On the Improvement of Understanding,* "The ideas which we form as clear and distinct, seem so to follow from the sole necessity of our nature, that they appear to depend absolutely on our sole power; with confused ideas the contrary is the case. They are often formed against our will." Today, teachers are rendered virtually powerless, and the ideas of the classroom are often not their own. The impositions of standards and mandates defeat the purpose of education. These school committees leave teachers *and students* with little understanding and control.

Interestingly, Eisner (2006, 46) concludes with an acknowledgment that "the satisfactions of teaching extend beyond the academic…indeed, the most lasting contributions come from saving lives, rescuing a child from despair, restoring a sense of hope, soothing discomfort. We remember these occasions longest because they matter." I do not read anywhere in the standards promulgated by a single state agency an objective that aims to achieve a single one of these most lasting contributions. They have clearly nothing to do with the annual yearly progress of any single school, and because they are immeasurable, they are also immaterial.

I am myself not inclined to listen to the pratings and beratings of these school committees.[8] The lawyer tells Bartleby that he ought to look at the green grass growing within the prison walls, but Bartleby responds, "I know where I am." Our school committees tell us we are free and treat us as enslaved; they tell us to speak and give us what to say. But, I know where I am. Here I say, "I want to make an extreme statement, if so I may make an emphatic one, for there are enough champions of civilization: the minister, the school-committee, and every one of you will take care of that" (225). And if Thoreau's extreme statement announced that it was in Wildness that the preservation of the world depended, then I wish to say here that it is in ethics that the preservation of pedagogy depends. And I would add that the six satisfactions of teaching noted by Professor Elliot Eisner depend *first* on the acknowledgment that teaching is a stance taken before the Other, and that everything that occurs in the classroom derives from that ethics. Teaching is a relationship founded on such ethics, and the satisfactions of teaching are the realization of an ethical stance. I would like to study the classroom from the basic question addressed by Eisner's piece: what are the ethics of the classroom that has impelled me to be a teacher, and to teach others to enter the teaching profession?

But perhaps I might start with the negative image of Eisner's positive; let me note a few factors of the teaching profession which would deter satisfaction in it.

Why should I teach? Eisner, a full-time professor at Stanford, does not mention money as a factor for choosing the teaching profession, and I presume he earns a salary far in excess of the majority of teachers in the United States today. However, many of those who drop out of teaching do so because of the lack of adequate pay. The AFT teacher salary survey (http:// www.aft.org/salary/index.htm) found that the average teacher salary in the 2003–2004 school year was $46,597, a 2.2 percent increase from the year before. However, even with the increase, teacher salaries fell short of the 2.7 percent rate of inflation for 2004. Interestingly enough, the United States Census Bureau (http://www.alaska.edu/swoir/surveys/index_docs/ incomechart.pdf) reports that the average family income in 2001 based in the educational attainment of the householder was $110,000 for those with a master's degree. Half of all public school teachers have earned this degree. Nonetheless, their salaries in 2005 fall well below the average for wage earners of comparable educational attainment. In addition, many states are attempting to drastically reduce or eliminate pension and health-care benefits, which were once negotiated as part of teacher compensation. Furthermore, in many states, teacher strikes are illegal, and thus teachers are powerless to change their own salary levels.

As for myself, the pay is generally abominable, though certainly service-able. I make a sufficient salary, though some might think me extravagant. Then again, I have two teen-aged daughters who love to shop, and I appre-ciate the spending of money myself. I used to feel quite guilty about my economy, but Max Horkheimer suggests that consumption, albeit conspic-uous, can be understood as a revolt against the American Puritan strain of asceticism and denial. It comforts me to know that when I saunter through the Mall of America I am striking out against the ethics of somberness and austerity prevalent in our own homegrown Calvinism. And though I don't own a Lexus, I can afford to change the muffler every so often and drive quietly down the streets of my hometown, silently subverting the zeitgeist. I do not teach for the money. Once, Rabbi Hayyim of Krosno, a disciple of the Baal Shem Tov, was watching together with his disciples a tight ropewalker. He seemed so absorbed in the spectacle that his students asked him what it was that riveted his gaze to this somewhat senseless per-formance. "This man," Rabbi Hayyim responded, "is risking his life, and I cannot say why. But I am quite sure that while he is walking the rope, he is not thinking of the fact that he is earning a hundred gulden by what he is doing, for if he did, he would fall" (Buber, 1975, 174). Perhaps if I thought about the money I would not teach; perhaps if I thought about the money, I could not teach. And of course, there isn't that much money about which to think.

Why should I teach? Certainly it is not for the status that attaches to the profession. Daily I read of my incompetence and that of my colleagues, and daily I am subject to the whims and fancies of our politicians who would transfer all of their incompetence over to us. If I were to believe even half of what they say about us, I would on most days call in sick, for indeed, that is what they say I am. William Bennett, former secretary of education, in a speech in 2000 honoring the Heritage Foundation said that "In America today, the longer kids stay in school the dumber they get relative to students in other industrialized countries" (Bracey, 2003, 67). My colleagues and I are Bennett's target. As my grandmother might say, *Vey iz mir!* Gerald Bracey writes (106) that "Our schools have been assailed decade after decade...many within the field of education have shown minimal support for public schools." We are a beleaguered population beset upon by hostile forces. We are constantly wrong and in need of correction. A state official, who should be *serving* the teaching professional by asking what he might do to enhance our work, rather than assuming our incompetence by directing its practice, spoke at a meeting outlining yet another set of newly mandated state revisions of the teacher education programs to which we are regularly subject. I asked him what evidence existed that previous to these newly promulgated programs we had not been doing our work with good results. He stuttered and blustered and ignored me. We *had* been doing our work with good results, but without these new mandates how would the bureaucrats justify their own positions?

When Rabbi Levi Yitzhak became the Rav in Berditchev, he made an agreement with the town leaders: they were not to ask him to their meetings unless they intended to discuss the introduction of a new usage or a new procedure. One day, they invited him to a meeting, and Rabbi Levi Yitzhah arrived expectantly. "And what is your new procedure or usage?" he asked. And the leaders described the institution of the *tzedakah* box into which the rich would put monies for the sake of the poor. When he heard what they had instituted, Rabbi Levi Yitzhak scolded them saying, "Did I not tell you not to take me away from my studies for the sake of an old procedure or usage." But the leaders protested that they had, indeed, invented something wholly new. And Rabbi Levi Yitzhah reminded them that the practice of not looking the needy in the face was as old as Sodom, where a young girl was punished for handing a beggar a piece of bread face to face instead of placing that bread in the charity box.

The revisions in our program were arduously made and kept us from our studies. In the newspapers they still condemn our practice.

Why should I teach? Eisner says that to teach is to make a difference. He says that immortality is the teacher's reward: "living in the memories of our students is no meager accomplishment" (44). Perhaps this may be so, but as teachers we have no control over the character and quality of those memories, and as our standards-based curriculum comes to dominate our pedagogy, the distinction between one teacher and the next disappears. My friends tell me that to teach is to touch minds, to teach is to change lives, to teach is to stimulate thought and be open to the ideas of our students. It is true that students have been affected by my pedagogy; I have met them on the streets and received communication from them after long absences and silences. Some of them I could not tolerate when we shared the classroom. Some of them remember a completely different classroom and pedagogy. I knew not what I did. Immortality, of which I must later have a great deal more to say, is a dubious satisfaction: it exists only in my hope and is almost wholly unrealizable.

Of course, on the doctoral level, or at what are called "the elite schools," of which Eisner's Stanford is certainly one, this relationship between teacher and student, and the possibility of making differences may be much more palpable and immediate; our students there regularly sit in our offices, meet with us for coffee and conversation, come to our homes for dinners and soirées. There, our graduate (and even undergraduate) students join us by choice, though they do not always choose *us*. They sit in our classrooms and at least pretend to attend. We swim in the heady world of meaning. A story has been told: It has been said that John Dewey would lecture while staring out of the classroom window overlooking the campus and surroundings of Teachers College. At the end of the hour he would turn to the class and say, "Thank you all. I think everything is a bit clearer now!" In this tale, students go off to discuss, to ponder, to question the matter of that classroom, and John Dewey goes back to his study to continue his work. Or he returns to his office to meet with interested students.

But there are too few of us privileged to teach these students. Here, in the public elementary and secondary schools, there are presently fifty million students and they attend not by choice but by law. In the public schools, at the hour's end,[9] teachers and students head off to the next class amidst great noise and clatter and confusion, and the heady world of ideas falls, like Icarus in Breughel's famous painting, fatally unnoticed and without alarm. Indeed, classroom management procedures we are urged to inculcate in our teacher education classes warn against just such teacher inattention enjoyed by John Dewey. He would effect education—indeed, there are few I know of greater influence than he—but in that classroom he seems unconcerned with affecting students.

Many of us, however, work in the public schools—there are six million teachers in the United States—and I think that if each of us touched only a *single* mind, how might the world be changed! Alas, I see in the world little evidence of that now, but regardless of the charges, we are not blameworthy. My students tend to just sit there and wait for my pronouncements because they are terrified of the classrooms where we, who are daily judged, sit in judgment on them. They are terrified students. They will enter classrooms filled with terrified students. I would not learn their terror. The Rabbis say, "One who learns from the young, to what can he be compared? To one who eats grapes that are tart, and drinks wine straight from the vat. But one who learns from the old, to what can he be compared? To one who eats grapes that are ripe, and drinks wine that is aged" (*Pirke Avot*, 4:26). Alas, I have eaten too many tart grapes, and drunk too long from the vats. I would sit down with the old. I am myself getting on.

Though, of course, I am being just a bit specious and dramatic. As Rabbi Hanina says, "I have learned much from my teachers, and from my colleagues more than from my teachers, and from my students more than from all of them." I remember once a student approached me after class somewhat apologetically. He shifted his slide rule back to his left side, and said he had a favor to request. It seems his application for the Westinghouse Science Scholarship was in the typewriter, and if he removed the paper to type up his essay for my class, he would never be able to realign the original page for the scholarship to continue typing it. He requested an extension on the paper that I had assigned. These were the days before computers, and even in wealthy Great Neck, most homes had only one typewriter. I asked him what his Westinghouse paper was about; he proceeded to name the title for his project, and I knew immediately that I didn't have the slightest idea about what he was talking. I probably could have learned a great deal then, but I had to get to my next class in a room three halls down. I agreed, of course, to the extension he requested. I do not think he won the award, but he did get an A on the paper. I drank to his success from the vat.

The problem remains that in the present climate little serious learning can take place. Students wait patiently for the information I am required to deliver, and patiently, I await the return of that information on the standardized and requisite exams. I am told that each student must write a philosophy of education, but when I ask whether students will be required to take a philosophy course so that they might understand what philosophy is about, I am told that there is no time for such nonsense. There are classrooms that require management! Students desire only to know what will be on the test, because in the present climate only that which is on the test counts as knowledge. What I learn from my students today is how educational policy today crushes the desire for learning.

Why should I teach? Eisner suggests that to teach is to engage in the world of great ideas. This smacks to me of Socrates' marketplace, where the wise man stood so engrossed in thought that he did not move. His students gathered about him. My students, however, move at the clanging of the bell. They come late to class unprepared. They ask what will be on the test. For ideas they care little.

In that classroom in which one of the satisfactions resides in the ability to introduce to students great ideas on which they may reflect for the rest of their lives, Eisner (2006, 44) valorizes the question. "Great teaching traffics in enduring puzzlements, persistent dilemmas, complex conundrums, enigmatic paradoxes." Great teaching eschews the answer. Eisner exclaims, "Questions invite you in. They stimulate possibilities. They give you a ride. And the best ones are those that tickle the intellect and resist resolution" (44). But the reality of the current ideological climate mandates a curriculum that is materially assessable. There must be only certainty and answers in this teaching. Today, in our classrooms, curriculum standards ensure that we have little opportunity to travel on the heady highway of ideas or to revel in the question. In fact, it is the answer upon which we insist, and we punish the question. If, as Eisner insists, great teaching pursues the question, then in schools today we remain quite sedentary. "I have met with but one or two persons," Thoreau (2001, 235) bemoans, "in the course of my life who understood the art of Walking, that is, of taking walks..." Thoreau assumes that those who do not walk have already taken to their beds; they do not saunter. Our classrooms are too often filled with those asleep.

It is flattering to assume that the ideas I transmit are great, but it is hubris to think that I convey to students those ideas. One of Spinoza's admirers, and perhaps a student of his, a certain Albert Burgh (1955, 410, 413), writes to him: "Even as formerly I admire you for the subtlety and keenness of your natural gifts, so now do I bewail and deplore you... Will you, wretched pigmy, vile worm of the earth, yea, ashes, food of worms, will you in your unspeakable blasphemy, dare to put yourself before the incarnate, infinite wisdom of the Eternal Father?" Love? His affections do not this way tend. No doubt Spinoza thought that he had trafficked in great ideas, even walked with his companion out of the village and into the woods; indeed, even dear Alfred seemed to have recognized his sauntering in the heady world of Spinoza's ideas. But the walk had come to a crushing halt, and Spinoza's companion became a dangerous stalker. Spinoza (1955, 415) addresses Albert calmly but with alarm when he regrets that his student has apparently abandoned everything he has previously learned and accepted. "I have thus been induced to write you this short reply, which I earnestly beg you will think worthy of calm perusal." Alas, it must be terribly frustrating to discover that our students have chosen a walk that contradicts everything in which we believe, by which we live, and about which we thought we have taught. It is small comfort, *very* small comfort indeed, to consider that our students know enough to reject us, and it is no comfort to know that they have rejected us to choose exactly that which is diametrically opposed to everything for which we stand, to opt for what Spinoza calls "superstition." As the poet so clearly stated, "There is no joy in Mudville, Mighty Casey has struck out!"

Well, perhaps, I teach for the learning. But what learning might mean is not an uncontentious issue. A story is told:

> In the dark, stuffiness of the cheder, the Rabbi stands before the class of children. The youthful charges, some not more than five or six, sit on the benches with their feet dangling several inches from the floor. The Rabbi

is tall and his beard is wild and flowing. He wears the traditional garb: black trousers and what seems to be a permanently soiled white shirt, black coat which flies up behind him as he paces back and forth before the students, and a black *kippah* (skullcap). Thee students sit silently, expectantly and terrified. He stares at his charges and in a stern voice asks, "Class, can any one tell me why the first book of the Bible, *Bereshit*, [ברשית] begins with the letter *gimel* [ג]?" Confused and panicked, not a child moves, nor offers a response. Each sits with his head down trying desperately to avoid notice. "Well," the Rabbi thunders, "can anyone tell me why the first book of the Bible, *Bereshit*, begins with the letter *gimel?*" He stares with wild, almost crazed eyes at the frightened children sitting quaking in their seats. The silence roars. "Can no one give me an answer?" the Rabbi roars. Suddenly, from the back of the room a hand goes up. "Yes," the Rabbi says, "can you please tell me why the first book of the Bible, *Bereshit*, begins with the letter *gimel?*" In a hushed and tentative voice, the child responds, "But Rabbi, the first book of the Bible doesn't begin with the letter *gimel* [ג]; it begins with the letter *bet* [ב]" The other children sit horrified and await tremblingly for the wrath of the Rabbi to come down on their poor colleague. The Rabbi stares out at the brave child for another ten seconds, and he says with a soft lilt in his voice, "Well, that's one answer!"

I love that story. In that bleak and dark classroom, conversation would never end. Yes, I would respond, that is *one* answer. Can someone offer another? For every answer, another question. There is, of course, a danger for me in that open classroom: If I teach students how to ask questions, then how can I stop them from asking me questions for which I have no answers? How shall I control the classroom when the world might be in there? And if answers are not readily forthcoming, can I consider myself yet a teacher? What should occur in the classroom that learning would take place? What is the learning that must take place in the classroom? What is the motive for a type of learning based in objectives? But no, that is a false question because I already know the answer. It is the indoctrination that maintains our slavery.

Present issues of curriculum swirl also about issues of control: control of children, control of teachers, control of knowledge, and control of the dreams. Control—that is today the only answer admissible in the classrooms of the United States. I suspect that it is comforting for some to know exactly what material must be covered; it must be satisfying to know that the material has been covered. But I am not certain that such teaching involves learning. Couldn't anyone else do that but me, and leave me in the marketplace with my ideas? Linda Darling-Hammond (2006, 301) notes that "[M]any laypeople and a large share of policy makers hold the view that almost anyone can teach reasonably well—that entering teaching requires, at most, knowing something about a subject, and the rest of the fairly simple 'tricks of the trade' can be picked up on the job." These lay people are absolutely wrong!

Why should I be a teacher? Perhaps we might suggest that the satisfactions of teaching are to be achieved in its difficulties, and that these satisfactions are hard wrung. Standing ethically before our students, commanding them to command us, we demand their attention with our devotions. We demand they learn to be attentive. To teach is to assume an ethical position in an immoral world. To teach is to be a prophet in a degraded world. To teach is to not suffer silently, but to suffer nonetheless. To teach is to change the world student by student and paper by paper.

Chapter 2

A Pedagogue for Two Teachers

I am finishing my sixth decade. Every third thought is of the grade.

I have wondered in the previous chapter what remains today of motives for being a teacher. In education we function in harsh times, and the pressures simply increase as the rewards become increasingly elusive. Who would such fardels bear? And I have been also wondering if there has ever been a time when education didn't exist in a sort of state of siege. At a conference presentation decrying the present acquiescence of teachers to the bellowings of politicians and technocrats with regard to education in the United States, I not-so-innocently asked when ever had existed that Golden Age of American Education when *all* children were contentedly and meaningfully educated, and *all* teachers were satisfied with the quality and prior learning of their students, and no parent or bureaucrat bemoaned the illiteracies promoted in schools, and *no* politicians promulgated laws that destroyed the very essence of education. When was the time in the United States, I wondered, when the teacher was an object of respect and praise and the scholar a figure of some authority? Indeed, I wondered, where are the snows of yesteryear? I recall that Socrates was executed for his teaching. Though teachers in our schools and academies are not threatened with physical obliteration, we are regularly endangered with moral devastation. William F. Pinar (2004, 34) argues that we can hope for little support from the public. "Given our conception by others, we are currently unable, as individuals or as a group, to undertake radical reform." Schools of Education are vilified and the professors in them scorned and attacked. Gerald Bracey (2003, 106) writes that "Our schools have been assailed decade after decade... [even] many within the field of education have shown minimal support for public schools." Bracey notes that in 1993, Albert Shanker, the president of the American Federation of

Teachers, charged that the achievement of K-12 students in the United States was poor, that American students were performing at much lower levels than students in other industrialized nations, and that international exams meant to compare students around the world showed students in the United States at or near the bottom. Though none of these accusations were then or are now true, the assault continues, the walls weakened and the fields overrun. We are a beleaguered population beset upon by hostile forces. We are constantly wrong and in need of correction. We get no respect.

In my lifetime, at too many cocktail parties, when to the question posed I announced that I was a teacher, the response too often was "Oh, my wife (or daughter, or mother) is a teacher." This was never meant as anything but disparagement: teaching was a woman's job, a holding tank until marriage or pregnancy facilitated a movement out of the school. The classroom, it was held, was no place for a man. Indeed, as Washington Irving (1961, 334) notes in "The Legend of Sleepy Hollow," though the school teacher often supplements his paltry income by helping the farmers with 'lighter' chores and in this process relieves the citizen's burdensome cost of education, it is really in the company of women that the male teacher is most comfortable. "The schoolmaster is generally a man of some importance in the female circle of a rural neighborhood, being considered a kind of idle gentlemanlike personage, of a vastly superior taste and accomplishments to the rough country swains, and indeed, inferior in learning only to the parson." Alas, this particular pedagogue, Ichabod Crane, this ineffectual and effete dandy, whose appearance is very much like that of a crane, is run out of town by the fearsome Brom Van Brunt, Crane's rival for the hand of the lovely Katrina Van Tassel. Ichabod Crane "was tall, but exceedingly lank, with narrow shoulders, long arms and legs, hands that dangled a mile out of his sleeves, feet that might have served for shovels, and his whole frame most loosely hung together. His head was small, and flat at top, with huge ears, large green glassy eyes, and a long snipe nose, so that it looked like a weathercock, perched upon his spindle neck, to tell which way the wind blew" (332). Ichabod Crane, teacher, has no use value in Sleepy Hollow, nor is education much more than a nuisance to the town. With Crane's exit, having been run off at the end by the fearsome Headless Horsemen, "the school was removed to a different quarter of the hollow, and another pedagogue reigned in his stead" (358). As was so often the case, "the schoolhouse, being deserted, soon fell to decay, and was reported to be haunted by the ghost of the unfortunate pedagogue" (359). The incompetent and timid pedagogue, Ichabod Crane, is the laughingstock of the town and the brunt of its bullying and cruel pranks; it is the image of Ichabod Crane who remains for our culture the quintessential impotent male presence in the classroom, and the quintessential symbol for the ineffectiveness of the entirely gendered classroom.

But teaching, as I have said, must be viewed as an ethical stance taken in a world degraded by violence, greed and ignorance, and not the repository for society's misfits and incompetents. To teach is to believe in, and to daily work for, that better world. The teacher stands daily in the classroom not terrified as had been Ichabod Crane, but defiantly as Bartleby had stood in his prison and announces, "I know where I am," despite the lawyers' attempt to romanticize his situation. Bartleby had been imprisoned because he preferred not to engage, and his isolation from the world resulted from his refusal to participate in it. Bartleby's death is simply a continuance of his preference not to. The teacher, however, affirms her purpose every day by entering the classroom to offer whatever future the world makes possible. Not prepared to be run out of town by its bullies, the teacher premises her existence on her decision not to leave matters where they be but to demand change. The teacher rarely considers exiting the stage, but in this world chooses to "absent herself from felicity awhile, / And in this harsh world draw thy breath in pain, / to tell my story" (V.ii.332–333). Dylan sings, "Life is sad, life is a bust, all you can do is do what you must / You do what you must do, and you do it well." Such is, and has been, I believe, the teacher's stance in the classroom.

It is hard, actually, for the teacher to ever leave the classroom, even when he or she is not physically there. I apologize continuously to my own children because I cannot cease being a teacher simply by exiting the classroom setting. I bring not only the work home but the obligations and responsibilities as well. As I have said, teaching is a stance to be taken in the world, and not merely a position to be occupied for the daylight hours. I suspect that it is a burden on our children to have teachers as parents; it is hard to step aside from the ethical stance that teaching demands. Too often my children don't want to know what I think of the movie we have just screened, or the TV show they want to watch, or the fan magazine they would purchase. "You'll just ruin it," they complain. Alas, I agree.

Perhaps part of the animus toward teachers derives from their being all about us in society. The teacher's presence reminds the rest of us of our need for education, if only, alas, we were not too busy. Thoreau (1991, 363) accuses, "With respect to a true culture and manhood, we are essentially provincial still, not metropolitan,—We are provincial, because we do not find at home our standards,—because we do not worship truth, but the reflection of truth,—because we are warped and narrowed by an exclusive devotion to trade and commerce and manufactures and agriculture and the like, which are but means, and not the end." Teachers are those who remain uncommitted to the pursuit of money and power and fame.

Teachers by their very presence remind us how imperfect we must always be; perhaps, society actively resists that knowledge by attacking the teacher. As Thoreau says, "[W]here there is a lull of truth, an institution springs up" (365). We teachers work in those institutions, and we teachers are the libidinal cathexes for the insecurities and rage that rise out of not knowing. As the six foot four basketball player highlights my moderate height, so the teacher *by contrast* points out what can (and too often, must be) yet learned. Teachers pay with our dignity for our jobs. Outside of the classroom, we remain teachers and everyone's animus; but in the classroom we are teachers and the students' animus. Who would such fardels bear?

I have spent more than half of my life as a teacher in the classroom, and for too much of that time, I, and my assailed colleagues have borne the increasing contumelies heaped upon the schools and its teachers. In an impossible profession, we are mandated to do the impossible, and then harshly scolded for each attempt. Linda Darling-Hammond (2006, 301) writes that "[T]he realities of what it takes to teach in U.S. schools such that all children truly have an opportunity to learn are nearly overwhelming." Who would such fardels bears? Gerald Bracey (2003) writes, "When I was a kid, adults often counseled me to enjoy what they knowingly said were the best years of my life. What adult would dare tell that to a child panicked over possibly repeating third grade because of a low test score. And who among our children having such dreadful experiences in school would then ever want to consider a career as a...teacher?" (184). Schools more than ever have become terrifying locations for those who must study and to work in them. Perhaps in these times a more productive question to consider is this: why refuse to be a teacher, to say, as it were, "I prefer not to." There are at least two interesting historical precedents for such a choice: Henry David Thoreau, whose words opened this chapter, and seventeenth-century philosopher, Baruch de Spinoza, whose words will end it. Baruch Spinoza and Henry David Thoreau refused the classroom as the venue for their work, perhaps for exactly this reason: they felt incapable of functioning there. Nonetheless, both devoted their lives to being teachers. I think that there might be something to learn about the classroom from the choice of each man to absent himself from it. Perhaps their refusals offer some insight into the present nightmare of the present, and some hope for the future. Separated by centuries and miles, my two excellent teachers speak in remarkable consonance decrying and refusing the life of the pedagogue. To me the patterns of their lives appear holographic: one reads the whole in each part.

In a letter dated February 16, 1673, Lewis Fabritius, Professor of the Academy of Heidelberg, and Councilor of Karl Ludwig, the Elector Palatine,[1] wrote to Baruch de Spinoza offering him on behalf of the elector "an ordinary professorship of Philosophy in his illustrious university" (Spinoza, 1955, 373). The invitation seems to have had no foreground: Spinoza had no university experience as either student or teacher, nor had he ever voiced any interest in securing a public professorship. Indeed, following his excommunication from the Jewish community in Amsterdam and his removal from that city in 1656, Spinoza had lived a life of relative quiet and seclusion, receiving many visitors, but apparently traveling very little, and even then only within a small circle in the Netherlands: from Amsterdam to Rinsburg, then to Voorburg, and finally to the Hague, where he died in 1677. The exact motives for his itinerancy are not clear to me. Kayser and Einstein (1946, 145) suggest that Spinoza sought "a life of quiet." They write, tinged with not a little romanticism I think, that, "[H]e could be an observer of the stormy times. It was as if he were looking out of his quiet life through a hole in the wall, over into the confusion, passions and struggles that menaced the freedom of the country" (186). They suggest the world came to his door, but that he was content to remain often behind it. It is also true that excommunicated from his community in 1656, unwilling to subscribe to the Christian faith, and excoriated by almost all for his heretical views, Spinoza sought physical safety in the rural settings. In his solitude he gained peace and some indemnity. Perhaps he sought the perfect quiet.

It interests me to consider me why once settled following his original expulsion from the Jewish community in Amsterdam, Spinoza chose to move at all. Reb Dylan (2001, "Mississippi") writes, "Everybody's moving, if they ain't already there / Everybody's got to move somewhere." Perhaps without movement there is no life. But traveling doesn't always require moving great distances, and I suspect that this is what Dylan means. Thoreau (1970) urged his citizens to move, to "be a Columbus to whole new continents and worlds within you, opening new channels, not of trade, but of thought" (438); I do not think Thoreau suggested by movement actual physical change of location, but rather of perspective. Possessions and permanence were anathema to Spinoza, too, it would seem, for whom these encumbrances were obstacles to exploration and freedom. Perhaps this might partly explain his peripatetic movement. Spinoza (1955) writes in "On the Improvement of Understanding," "All the objects pursued by the multitude not only bring no remedy that tends to preserve our being, but even act as hindrances, causing the death not seldom of those who possess them, and always of those who are possessed by them" (5). Spinoza (1970) foreswore the pursuit of material gain. Thoreau, too, complained of the burdens the material makes on the spiritual: "I had three pieces of limestone on my desk, but I was terrified to find that they required to be dusted daily, when the furniture of my mind was all undusted still, and I threw them out of the window in disgust" (174). Certainly, Spinoza's itinerancy had motives other than the quest for material aggrandizement. Indeed, Giles Deleuze (1988, 9) writes, "What defines Spinoza as a traveler is not the distances he covers but rather his inclination to stay in boarding houses, his lack of attachment, of possessions and property, after his renunciation of the paternal inheritance." Perhaps a certain transitoriness was what Spinoza sought, for in that impermanence he found a freedom, the inevitably failed consummation to which he devoted his life.

Other than his lens-grinding equipment, Spinoza owned very little, a fact, I suspect, which made movement even easier. Thoreau (1970) says, "Pray, for what do we *move* ever but to get rid of our furniture, our *exuviæ*..." (200), and Spinoza's movement precluded object accumulation. After Spinoza's death, his stepsister and nephew thought to inherit those things of value, but the total value of his possessions was so small that after the payment of his debts nothing would be left for them to inherit, and so they renounced their claim to Spinoza's things. Thoreau argued that a man is rich insofar as in the event of a fire, he can leave his burning home carrying on his back everything he owns. By this definition, Spinoza and Thoreau were very wealthy, indeed. If the cost of a thing was assessed at how much life it would require to possess it, then a permanent domicile was too heavy a cost for either Spinoza or Thoreau. Neither seemed willing to be so encumbered; they had more valuable places to dwell. Spinoza (1955) said, "...after I had recognized that the acquisition of wealthy, sensual pleasure or fame, is only a hindrance, so long as they are sought as ends not as means; [but] if they be sought as means, they will be under restraint, and far from being hindrances, will further not a little the end for which they are sought..." (7). Those possessions sought as ends deter us from thought and freedom. Thoreau (1970) said, "It is possible to invent a house still more convenient and luxurious than we have, which yet all would admit that man could not afford to pay for" (173). For both Spinoza and Thoreau, I think, it was the life of the mind and spirit that they sought forswearing all that might physically root them in place; each sought a freedom he could carry with him.

If Spinoza could be said to be a traveler at all, it was in his mind that he journeyed, and there he appeared to move extensively and continuously and productively. It was in his thinking that he was most at home. In "On the Improvement of the Understanding," Spinoza (1955) writes, "This, then, is the end for which I strive, to attain to such a character myself, and to endeavour that many should attain to it with me. In other words, it is part of my happiness to lend a helping hand, that many others may understand even as I do, that their understanding and desire may entirely agree with my own" (6). I think Spinoza sought to teach others the freedom of travel. Spinoza seems to have also been surrounded by a far-flung social circle, and like Thoreau, was most inclined "to fasten myself like a bloodsucker for the time to any full-blooded man that comes in my way" (*Walden*, 271). Indeed, the incursion of strangers at his door might explain his first move to Voorburg. Kayser and Einstein (1946, 145) say that in Rinsburg, "quite a number of strangers did come [to his home], and this disturbed him. Rinsburg, which was the principal seat of the Collegiants, brought him all too near to the sectarians, who were quite distant and strange to him...He moved to Voorburg, near the Hague, where he had friends whom he could meet whenever he wished. Inquisitive and importunate people could not find their way to him; but his friends knew where to get in touch with him." In his small and periodic moves, Spinoza seems almost to have sought to establish a place to which sympathetic minds could come for conversation and learning, and even some simple refreshments. Kayser and Einstein (164) quote an unnamed source: "All his activities were now focused on how he could actively help his neighbor, and how he could avoid becoming burdensome by ill temper either to his neighbor or to himself." I think in his purposes Spinoza established an alternative school where "the goal of knowledge should not be separated from our conduct of life...Spinoza, the most solitary of souls, also was a preacher of social assistance" (164). Spinoza was a teacher.

Though Spinoza's surviving correspondence might appear relatively small, he engaged in the practice of letter writing enthusiastically and with great attention, explaining and clarifying his views to whomever cared to inquire. For example, in an extended series of letters, Spinoza engages with Hugo Boxel in a spirited discussion concerning the existence of ghosts. As I had come to expect, Spinoza treats his student with respect and a skepticism inspiring further thought and conversation. More, Spinoza (1955) acknowledges that, indeed, there are more things in heaven and earth than are dreamt of in his philosophy, and so he is prepared to consider, though later reasonably refute, the existence of ghosts. "If philosophers choose to call things which we do not know 'ghosts,' I shall not deny the existence of such, for there are an infinity of things, which I cannot make out" (377). As the best of teachers, Spinoza is never afraid to admit that he does not know. In this case, however, very logically and rationally, Spinoza argues against the reality of ghosts with his student. It was not seclusion Spinoza sought, but solitude.

This desire to improve the lives of their fellows defines Spinoza and Thoreau as both student and as teacher. It is not conformity of thought that either desired; rather, both sought a freedom for all that *derives* from an ability to think freely and desire freely. Spinoza writes in

A Theologico-Political Treatise (2005, 202), "When we reflect that men without mutual help, or the aid of reason, must needs live miserably, as we clearly proved in Chap. V, we shall plainly see that men must necessarily come to an agreement to live together as securely and well as possible if they are to enjoy as a whole the rights which naturally belong to them as individuals, and their life should be no more conditioned by the force and desire of individual, but by the power and will of the whole body." Spinoza advocated for the establishment of democracy, for only in a democracy could humans be free enough to come to understand "the external and internal factors governing their lives, and thus cease to be dependent on those forces. Rather, with understanding we take control over ourselves and assume responsibility for our actions. It is to this end that Spinoza endeavored throughout his brief life. Spinoza desired that learning should serve one aim: '…that we may attain to the supreme human perfection' (7). For Spinoza, these perfections consisted of the improvement of the understanding, 'the capacity to apprehend things without error, and in the best possible way' (7). Smith (2003, 26) writes that the *Ethics* teaches that while we cannot escape nature, and that much that goes on within it will forever elude human control, we can take responsibility for our lives and how we choose to live them…Spinoza teaches us not only to take responsibility for our lives but to find joy and happiness in doing so…He makes the joy of life itself his greatest good." Well aware that as a finite human being he must fail at attaining this perfection, Spinoza did not cease working toward it. Though he acknowledges himself that "the things which I have been able to know by this kind of knowledge are as yet very few," Spinoza believed, like the Rabbis whom he had studied so well, "It is not your responsibility to finish the task, yet you are not free to withdraw from it" (*Pirke Avot,* 2:21).

Both Spinoza and Thoreau taught that life was about method and not content. Method, Spinoza (1955) said, "is that which teaches us to direct our mind according to the standard of the given true idea" (16). Spinoza sought to achieve the stable human character that possesses the "knowledge of the union existing between the mind and the whole of nature" (6). In a like manner, Thoreau exhorts: "[If] one advances confidently in the direction of his dreams, and endeavors to live the life which he has imagined, he will meet with a success unexpected in common hours. He will put some things behind, will pass an invisible boundary; new, universal, and more liberal laws will begin to establish themselves around and within him, and interpreted in his favor in a more liberal sense, and he will live with the license of a higher order of beings. In proportion as he simplifies his life, the laws of the universe will appear less complex, and solitude will not be solitude, nor poverty poverty, nor weakness weakness" (440). Both men sought to improve the understanding for themselves and for their fellows: for what else do we educate, and for what other purpose should a teacher serve but as provocateur and guide.

Spinoza apparently enjoyed a very simple, rather mundane life, ate gruel on most days, drank a little beer and infrequently wine, and regularly smoked a long-stemmed clay pipe as was the custom in seventeenth-century Holland. His first biographer, Colerus, says, "It is scarce credible how sober and frugal he was. Not that he was reduced to great poverty, as not to be able to spend more, if he had been willing. He had friends enough, who offered him their purses, but he was naturally very sober, and would be satisfied with little" (Spinoza, 2005, xviii). However, Spinoza was not a rich man, and perhaps the salary from the teaching position would have permitted him a bit more in the way of material comforts. Spinoza was, however, wary of and disinterested in such financial ties. He (1955) urges that we ought "To endeavour to obtain only sufficient money or other commodities to enable us to preserve our life and health, and to follow such general customs as are consistent with our purpose" (7). That purpose, of course, is the improvement of our understanding and the achievement of a satisfying and fulfilled life. Simon Joossten DeVries, a close friend of Spinoza, tried to give the philosopher a gift of two thousand florins, "so as to enable him to live more comfortably," but Spinoza refused to accept the money. When DeVries died, he wanted to make Spinoza his sole heir, but Spinoza thought that this would be unfair to DeVries' brother and he refused it, so the brother decided to give Spinoza an annual stipend of five hundred guilders. Spinoza thought this figure excessive, and insisted that it be reduced to three hundred guilders. In his biography of Spinoza, Jean Maximilian Lucas, a French Protestant refugee living in Holland, stated that "He cared little for the goods of fortune... Not only did riches not tempt him, but he even did not at all fear the odious consequences of poverty" (Nadler, 1999, 262). No, it would not be the salary that would induce Spinoza to accept the position offered him by His Most Excellency.

Thoreau, too, lived a life of solitariness amidst great sociability. "A man thinking or working is always alone, let him be where he will," he announces boldly (267). He decries the current insistence on society and sociability, finding too much company distracting. "We are the subjects of an experiment which is not a little interesting to me. Can we not do without the society of gossips a little while under these circumstances—have our own thoughts to cheer us?" (266). Thoreau, too, sought the uncluttered life. "Simplify, simplify, simplify," he admonished. For Thoreau, as for Spinoza, the pursuit of wealth and the acquisition of material possession was, in fact, an ephemeral distraction from the real purpose of life: "Superfluous wealth can buy superfluities only. Money is not required to buy one necessary of the soul" (444). Freedom can only be earned and not purchased.

The offer of an academic position had been made nonetheless to Spinoza, and the terms appeared ostensibly quite generous. "An annual salary would be paid to you, equal to that enjoyed at present by the ordinary professors. You will hardly find elsewhere a prince more favourable to distinguished talents, among which he reckons yourself. You will have the most ample freedom in philosophical teaching, which the prince is confident you will not misuse, to disturb the religion publicly established." It took Spinoza a little over a month to graciously refuse the offer.

Fabritius' is a curious caveat, since the very charges leveled at Spinoza accused him of exactly that which in his offer the Elector Palatine expressly forbade: disturbing not only the religion publicly established, but, in fact, condemning all organized religions regardless of establishment. Thus, this particular offer of a professor's chair to Spinoza by the elector palatine appears disingenuous at the least, and dangerous at best. It seems almost to me as if the offer was intended not to provide Spinoza with a safe and public forum, but to prevent him from continuing his work. The warning in the offer of the professorship appears a way to silence Spinoza and keep him in check, or better, to get him to do work of which the state approved, an ironic acknowledgment of his ability to influence. Spinoza saw through the ruse and politely preferred not to.

Today's offers of the classroom to teacher-educators seem as specious and dangerous as Fabritius' offer of an ordinary professorship to Spinoza. The high stakes testing movement, exacerbated by the No Child Left Behind (NCLB) legislation, holds the sword of Damocles over the head of each and every school teacher in the United States, and threatens each with severe consequences for disturbing the publicly espoused religion. Gerald Bracey (2006, 106) writes that "In the current climate, and under the No Child Left Behind law, teachers and administrators can have their competence judged by test scores and can lose their jobs if the test scores come up short. Under state laws, teachers and administrators might get promoted or terminated because of the test scores." How this religious principle evolved is not my subject for discussion here, though I am not uninterested in this development. What I am interested in, however, is to what extent these caveats deter the best and the brightest from entering and remaining in the profession. Perhaps Spinoza might cast some light on it.

In the *Tractatus* Spinoza (2005, 206) says, "Action in obedience to orders does take away freedom in a certain sense, but it does not therefore, make a man a slave, all depends on the object of the action. If the object of the action be the good of the state, and not the good of the agent, the latter is a slave and does himself no good." As Hamlet says, "That would be scanned." If the state mandates that every child will take a test *every year* to measure the degree of her learning, then it is clear that the state also mandates that every teacher prepare students to take the test by teaching the materials that will be tested. The curriculum must focus on those materials so that the child's annual yearly progress may be properly measured. The sum total of the annual yearly progress of the students in a particular school measures the annual yearly progress of the school, measures the educational value of that school, measures the competence of the teachers in that school. No other contingent factor in considering a school's work may be taken into account outside the absolute numbers on the tests. Any curricular matters that do not address standards to be tested are deemed irrelevant, and any teacher who neglects these standards is declared incompetent. It is clear that the teacher is not free to do anything other than what the state mandates. The teacher, far from having the freedom of action, and being assumed skilled and competent, becomes the state's slave. The teacher, far from having the opportunity to traffic in enduring puzzlements must wallow in the swamps of facts and numbers and tests. In Spinoza's words, the teacher is a slave and does himself or herself no good. Who would such fardels bear to grunt and sweat under a weary life?

If, as Mary Poppins says, a spoon full of sugar helps the medicine go down, in this system the teachers are now defined as the bearers of, even identified sometimes as the substances of, such sweetness as to ensure the ingestion of the mandated medicine. Or not to swallow it, as the case may be.[2] But regardless of point of view, whether the teachers bear sweetness or acerbity, deliver the medicine they must. Little children who get caught out in a rainstorm without wearing their galoshes, Mary Poppins warns, must take their medicine. There is, of course, some benefit if your nanny can make the medicine taste like, say, strawberry soda; nonetheless, the state mandate insists that for the child's own good the medicine must be taken. The state has determined what knowledge is of most worth, and teachers must teach this material, even as students must learn it. The state has turned teachers and students into slaves in the excesses of its own grasping for power. There is nothing about the testing that is good for the teacher: as Spinoza says, the teacher is a slave.[3]

Rather, it is the state bureaucracy in its voluble attack on the educational system, its charge of teacher incompetence, and its assumption of total control over the schools and the teacher's autonomy, which benefits from these actions. In the control of the educational apparatus, the state asserts its control over the future *and* the past. As Hamlet says, "Let me not think upon't." Spinoza (2005) says that regardless of whether "the act is done out of love or fear … every action which a subject performs in accordance with the commands of the sovereign, whether such action springs from love, or fear, or (as is more frequently the case) from hope and fear together, or from reverence compounded of fear and admiration, or, indeed, any motive whatever, is performed in virtue of his submission to the sovereign, and not in virtue of his own authority" (215). Thus, every teacher is transformed into a slave, and her work reduced to coerced labor. Spinoza suggests, however, that the free teacher must consider himself or herself so competent that the decisions in the classroom derive from their own expert reason, and that they need depend on no external authority other than their reason. It is on this authority that Spinoza (369) would say that anyone who wishes to teach should be permitted to do so, provided they are prepared to assume their own cost and risk! Alas, today, there is no motive for choosing teaching or the teacher's role, and the risks involved, externally imposed, are severe.

But it is not just the teacher whom the state enslaves. The high stakes testing has ensured that the child will learn only what the state requires to be learned for the good of the state, and for the sake of indoctrination denies to the individual child the freedom of learning. Bracey (2003, 183) writes that "The standards movement, alas, has returned to the conception that children are just small grownups, and the consequences could be severe. Throwing up on test day could be the least of it." Students suffer horribly from the testing and measurement regimen to which they are subjected, and the teachers are reduced to mere slavery in their daily practice. Thoreau (2001, 248) writes that "Many a poor-eyed student that I have heard of would grow faster, both intellectually and physically, if, instead of sitting up so very late, he honestly slumbered a fool's allowance." There is no good in No Child Left Behind, and certainly little honest learning. I think that Spinoza eschewed the life offered at the university for the intellectual life he could enjoy outside of it.

When Fabritius made the offer of an academic position at Heidelberg, Spinoza must have been occupied with writing *The Ethics*, the book many consider his masterpiece, though the incendiary nature of the material precluded its publication during his lifetime. Indeed, *The Ethics* was not published until after Spinoza's death, and then, to ensure the safety of those still alive, the book was issued without publisher's name or place of publication.[4] *The Ethics* has been called one of the most difficult books ever written; what makes the book so demanding is the geometric method of its structure. But on at least one level, this method is integral to Spinoza's theme: the liberation of human life from the enslavement of dogma and coercion.[5] Smith (2003, 200) writes, "The geometrical method of the *Ethics* represented for Spinoza precisely such a model of human self-making, a construction of the mind that could serve as an expression of human power and freedom." *The Ethics* represents Spinoza's ultimate hope for human possibility; it seems to be the intellectual gift of hope and love from a great teacher. Smith says "*The Ethics* is a celebration of life, of joy and laughter, of sociability and friendship. Spinoza's philosophy culminates not in the grim and remorseless recognition of necessity...but in the enjoyment of the pleasures of mind and body working together as a unified whole that helps to secure the conditions of human autonomy" (4). *The Ethics* was meant as Spinoza's (1955) ultimate statement, his magnum opus, concerning the achievement of a life, the end of which would be a peace, a contentedness and a sense of place in the world. "Whereas the wise man, in so far as he is regarded as such, is scarcely at all disturbed in spirit, but, being conscious of himself, and of God, and of things, by a certain eternal necessity, never ceases to be, but always possesses true acquiescence of spirit" (270). I think *The Ethics* represents the effort of a very great teacher, though at the time of the offer of the professorship, Spinoza had published very little, and what little he had written and that had reached the public ear was, for the most part, excoriated for its author's heretical views and for its advocacy of a freedom few monarchs and clergy were prepared to offer.

Spinoza published with little success and great infamy. The *Theological-Political Treatise* had appeared in 1669–1670; it had met with a response both enviable, in that this was a volume that some had even read,[6] and horrible, in that those who read it despised it volubly and with a passion almost unprecedented. The *Tractatus* was published anonymously in Latin, as was the later *The Ethics*, but many knew the work to be that of Baruch de Spinoza. Steven Nadler (1999, 295) writes that "The book reached the Dutch reading public in early 1670. The reaction, far and wide, was immediate, harsh, and unforgiving. Spinoza quickly became identified as an enemy—perhaps the enemy—of piety and religion. Some of his more extreme critics accused him of being an agent of Satan, perhaps even the Anti-Christ himself." The *Theological-Political Treatise* appeared just two years before the violent assassination of the deWitt brothers, Holland's republican leaders; Spinoza's advocacy of the deWitts and his advocacy of the liberty and freedom they espoused, aligned Spinoza with what were considered dangerous elements in Dutch society, and endangered his well-being, and even his life.

When Fabritius extended the offer of an academic position to Spinoza, then, the solitary philosopher had lived most of his public life in disrepute. It must be recalled at this time that in 1656, when Spinoza was but twenty-four years old, the Jewish community had pronounced a *cherem* on him, and he had been excommunicated for reasons upon which we can only speculate,[7] but must certainly have had something to do with his criticism of Jewish beliefs and practices, his redefinition of God, and his critique of Scriptures. These issues would be formally explored in the *Tractatus*. In that work, the philosopher meant to reestablish the world on a liberal and rational basis, providing the maximum degree of liberty for all of the citizenry. The tyranny of the religious infrastructure, he argued, had enslaved the populace, subjugateds and denied it the freedom his philosophy espoused. "The aim of the work as a whole is the liberation of the individual from bondage to superstition and ecclesiastical authority. Spinoza's ideal is the free or autonomous individual who uses reason to achieve mastery over the passions" (Smith, 1997, 17). The freedom that Spinoza advocated in the *Tractatus* was intolerable not only to the rulers, but to the cultural elite as well, even as this freedom was unimaginable to the populace for which Spinoza really had little affection but for whom he had, apparently, high hopes. Spinoza's work was so radical that his book had to be published anonymously and under the imprint of a fictitious publisher.

Nonetheless, I think that Spinoza's idea of "positive liberty," "according to which one can only be free in a self-governing community where every citizen participates in the creation of law" (Smith, 2003, 126), seems a precursor to John Dewey's insistence that democracy was essential for a proper education, but that only by education could democracy exist.[8] For Spinoza, as for Dewey, "only the intelligent man could be free, but only the free man could be intelligent" (Hampshire, 2005, 107). Thus, whatever facilitates the development of knowledge and intelligence is necessarily good for the individual and should be pursued; whatever obstructs knowledge and intelligence must be avoided. As Dewey devoted his entire intellectual life to decrying and exploding the persistence of dualisms as false ways of defining the world, so, too, did Spinoza believe that the primary dualism separating mind and body, "like all dualisms, inevitably leads to contradictions" (72). I do not know that Dewey and Spinoza have ever been formally linked in academic discourse, but I draw a great deal of comfort thinking of John Dewey as a closet Spinozist.

Both Dewey and Spinoza sought to develop the rational capabilities of the human being for the purpose of living a more free and satisfying life. Dewey's idea of education, a reconstruction of experience, as postmodern as it is modern, as psychoanalytic as it is materialist, was to enable the individual to understand the meaning—individual, social, and cultural—of experience. Dewey (1964, 273) writes,

> When trying, or experimenting, ceases to be blinded by impulse or custom, when it is guided by an aim and conducted by measure and method, it becomes reasonable—rational. When what we suffer from things, what we undergo at their hands, ceases to be a matter of chance circumstance, when it is transformed into a consequence of our own prior purposive endeavors, it becomes rationally significant—enlightening and instructive.

Dewey never eschewed rigorous intellectual endeavors; rather, he understood that meaningful learning occurs best in the pursuit of meaningful living. Similarly, for Spinoza, learning has purpose and meaning when it occurs in the pursuit of living. If Spinoza believed that our learning was always incomplete, nevertheless, he knew that the more we learned, the greater would be our understanding of the natural laws by which we functioned and by which we participated in the world. Genuine knowledge, for Spinoza, would be a sequence of ideas "each one of which follows logically from its predecessor" (Hampshire, 2005, 74). This knowledge would make us free. "A free man," Spinoza (1955) writes, "is one who lives under the guidance of reason, who is not led by fear, but who directly desires that which is good, in other words, who strives to act, to live, and to preserve his being on the basis of seeking his own true advantage; wherefore such an one thinks of nothing less that of death, but his wisdom is a meditation of life...I call free him who is led solely by reason" (232). I think Spinoza very much meant to mentor to this wisdom those with whom he met and talked; I suspect he actually lived at moments in this wisdom. I think that Spinoza would have been an excellent teacher. But I do not believe that his effort would have neither then nor now have much efficacy in the schools.

John Dewey, I think, held out a similar hope for education. Dewey believed that it was not knowledge that the schools must seek to develop, but the capacity for wisdom and responsibility. This effort required the acquisition of knowledge. Dewey writes in *Democracy and Education*:

> A person who is trained to consider his actions, to undertake them deliberately, is in so far forth disciplined. Add to this ability a power to endure in an intelligently chosen course in face of distraction, confusion, and difficulty, and you have the essence of discipline...Discipline is positive. To cow the spirit, to subdue inclination, to compel obedience, to mortify the flesh, to make a subordinate perform an uncongenial task—these things are or are not disciplinary according as they do or do not tend to the development of power to recognize what one is about and to persistence in accomplishment. (129)

The schools, Dewey warns, ought not to crush individual interests, but neither must the schools indulge them. Rather, the curriculum must engage the interests of the student so that he or she will develop *discipline*, by which concept Dewey meant a continuity of attention and endurance. He held that the "training of thought" released the individual from the present, enabled them to plan the future, and arranged for object use. Thought permitted untrammeled access to the world and gave power to the student. Dewey (1964, 400) writes that "Intellectual integrity, honesty, and sincerity are at bottom not matters of conscious purpose but of quality of active response. Their acquisition is fostered of course by conscious intent…" It is the process of thought that must be developed *in* the school and *through* the curriculum if education was to be effective for the advancement of democratic life.

Spinoza demands from "the wise man" the same rigor as Dewey demands from the school, and for much the same purpose. Spinoza understands that all errors are the result of a lack of knowledge, and that all errors are not absolute, but are products of incomplete knowledge. Learning is the antidote to ignorance, but as we are often reminded, "It don't come easy." Spinoza (1955, 270–271) writes at the end of the *Ethics:* "If the way which I have pointed out as leading to this result seems exceedingly hard, it may nevertheless be discovered. Needs must it be hard, since it is so seldom found. How would it be possible, if salvation were ready to our hand, and could without great labour be found, that it should be by almost all men neglected? But all things excellent are as difficult as they are rare." Dewey never felt that school must be made easy, but only that it should be meaningful. Dewey (1964) writes, "Only by wrestling with the conditions of the problem at first hand, seeking and finding his own way out, does [the student] think" (160). Spinoza never meant that education should be productive; rather, I believe he held that education should be redemptive. These two are my teachers, but if Dewey spent much of his life in the public classroom, Spinoza rejected this venue. Though the offer might have promised some public recognition of his work, though the offer might have offered him a greater audience for his ideas, though the offer might have ended his self-exile in the Netherlands, Spinoza rejected Fabritius' offer of an ordinary professorship.

In reality, it does not appear that Fabritius relished the task of making the invitation to Spinoza on behalf of Karl Ludwig: Fabritius despised the little he had seen or heard of Spinoza's work. He had apparently read the *Theological-Political Treatise*, and he expressed horror and outrage at its matter. He wrote to his friend and biographer, Joseph Heidegger, "I shudder when I see such unbridled licentiousness being presented in a public display, and the Christian religion itself and Holy Scripture being so openly blasphemed" (Nadler, 1999, 312). In fact, I do not think Fabritius cared to have any association with Spinoza or his ideas. Nonetheless, the Elector Palatine had directed him to invite Spinoza to join the university as a regular professor, and to offer him all of the amenities that attach to that position. As the elector's representative, Fabritius offered a valued position in the university to what would appear to be his intellectual nemesis.

In the medieval world, Heidelberg had been one of the premier institutions of learning, though by the time the position was offered to Spinoza, the intellectual and fiscal wealth of the university had considerably declined. Indeed, in 1622 the library of the university was stolen and removed to Rome! Karl Ludwig, the Elector Palatine desired to return Heidelberg to its former position, and perhaps it was this hope that provoked his invitation to Spinoza. But by the time of the Elector's offer, Heidelberg, as had many institutions, had become a Calvinist ideological bulwark, and heretical views were not easily tolerated. Spinoza was nothing if not a heretic. In 1656, as I have noted, he had been excommunicated from the Jewish community in Amsterdam for activities about which one could only speculate. Though he changed his name from the Hebrew *Baruch* to the Latinate *Benedict*, he never became a Christian, nor were the Christians at all enamored of his philosophy. The radical Collegiant sect befriended him, but Spinoza spent his entire life as the enemy of all organized religion, including the Calvinist Christianity that dominated the Netherlands. Steven Smith (1997, 25) suggests that "Spinoza attacks Judaism as an opening gambit to get at Christianity. His real target, however, was Christian sectarianism and intolerance, not Jewish particularity and exclusivity." In the *Treatise*, Spinoza "champions the freedom of thought and opinion and the toleration of religious heterodoxy, argues for the subordination of the clergy to the secular powers, and defends the independent use of reason as an inalienable human right" (in Smith, 1997, 10). Given this theological and religious agenda that represented no less than a revolution in medieval Christian worldview, it is no wonder that Spinoza aroused the hatred of those in power. Thus, the offer from Fabritius appears to be somewhat odd, even suspect. But it does seem to assume a great deal of power in the academic podium. Spinoza rejected the offer and the apparent power. That, too, must be scanned.

Spinoza clearly loved learning. He had been a star pupil in the Talmud Torah of Amsterdam, and, perhaps, if his father had not died, obliging Baruch to maintain with his uncle the family business, he might have studied to become a respected Rabbi in the community. His knowledge of languages included French, German, Italian, Portuguese, Spanish, and Hebrew. In addition to his work at Talmud Torah studying traditional Jewish texts and methods, Spinoza enrolled in the school taught by Franciscus Van den Enden, where apparently he learned Latin, studied philosophy and perhaps even acquired a political education, "not just in the sense that Van den Enden gave him the classic works of political thought to read (including Aristotle's *Politics*, Machiavelli's *The Prince* and the *Discourses*, Hobbes's *De Cive*, and Grotius's republican writings), but also that Spinoza's commitment to a secular, tolerant, democratic state was influenced both by his tutor's own opinions and by those whom he might have met in the house on the Singel" (Nadler, 1999, 113). It is probable that Spinoza assisted with the teaching at Van den Enden's school as well. Johannes Caesar became Spinoza's student "to obtain a thorough instruction in the Cartesian philosophy" (194), but this appears to be the extent of Spinoza's experience as a formal pedagogue.

Spinoza sought after an intellectual, if not an academic life, and perhaps thought of himself as a "philosopher." What this might have meant to him can be gleaned in part from a fragment published as part of his posthumous works. There Spinoza (2005) wrote:

> Philosophers conceive of the passions which harass us as vices into which men fall by their own fault, and therefore, generally deride, bewail, or blame them, or execrate them, if they wish to seem unusually pious. And so they think they are doing something wonderful, and reaching the pinnacle of learning, when they are clever enough to bestow manifold praise on such human nature as is nowhere to be found, and to make verbal attacks on that which, in fact, exists. For they conceived of men, not as they are, but as they themselves would like them to be. Whence it has come to pass that, instead of ethics, they have generally written satire, and that they have never conceived a theory of politics, which could be turned to use, but such as might be taken for a chimera, or might have been formed in Utopia, or in that golden age of the poets when, to be sure, there was least need of it." (Chapter I, Part I, p. 287)

It is not uninteresting to speculate that Spinoza believed almost two centuries *before* Marx that philosophers had yet only described the world, but that they had not yet undertaken to change it. I think Spinoza meant in his work to accomplish the latter. He meant *as a teacher* to offer to his fellow citizens a means to a life free of all that is not life. In another fragment, *On the Improvement of the Understanding*, Spinoza (1955, 3) sets out to learn, in a language Thoreau would approximate two centuries later, "whether, in fact, there might be anything of which the discovery and attainment would enable me to enjoy continuous, supreme, and unending happiness." I cannot think but that Spinoza's desire was to offer in his writings *and teachings* the opportunity of liberty and happiness to all. Spinoza's desire seems to accord with the satisfactions Eisner (2006) attributes to the classroom, but Spinoza refused the offer to enter it.

For Spinoza, life was a way of being and not a theoretical construct. In his uncompleted manuscript *On the Improvement of the Understanding* (1955) he offers an account of his own method and motivation: "One thing was evident, namely, that while my mind was employed with these thoughts it turned away from its former objects of desire, and seriously considered the search for a new principle; this state of things was a great comfort to me, for I perceived that the evils were not such as to resist all remedies" (5). What he learned and would teach is how to discover the good life and not what good life to live. A stranger to the land, a heretic excommunicated from his community, an exile in his own land, and yet a lover of liberty and the individual, Spinoza thought his way to freedom, and in *The Ethics*, sought to offer to his fellow citizens a means to achieve a life of freedom and joy. *The Ethics*, indeed, perhaps all of Spinoza's work, was an expression of absolute love for his fellows. Gilles Deleuze (1988, 13) says that the geometric method of *The Ethics* "is in opposition to what Spinoza calls satire; and satire is everything that takes pleasure in the powerlessness and distress of men, everything that feeds on accusations, on malice, on belittlement, on low interpretations, everything that breaks men's spirits..." The geometric method speaks not in hatred, but in love. I think Spinoza truly meant by his work to liberate his fellows whom he must have loved a great deal. I think he is a wonderful teacher.

Spinoza's reasons for declining the Elector's offer, then, are revealing; they speak to an integrity I had come in my own study to expect of him. In his refusal, he is so politic, so deferent, and so cleverly equivocal. He says, "If I had ever desired to take a professorship in any faculty, I could not have wished for any other than that which is offered to me, through you, by His Most Serene Highness the Elector Palatine, especially because of that freedom in philosophical teaching, which the most gracious prince is kind enough to grant, not to speak of the desire which I have long entertained, to live under the rule of a prince, whom all men admire for his wisdom" (Spinoza, 1955, 374). There is first the flattering suggestion that the Elector's offer is perhaps the only offer that Spinoza might have ever accepted, *if* he Spinoza had been ever inclined to accept a professorship in any faculty. Spinoza continues: "But since it has never been my wish to teach in public, I have been unable to induce myself to accept his splendid opportunity..." What a lovely qualification: it is not the particular offer that Spinoza declines, but the entire possibility of a teaching career that the offer represents. The refusal of the appointment by a man who openly espoused to change the world with his ideas, however, seems odd. To turn down the offer of such a public podium seems self-defeating. If Spinoza had no intention to teach, then what did he intend to *do* in writing such incendiary texts.

Teaching, Spinoza argues, would preclude the activity of philosophical research. Spinoza expresses concern that his teaching duties would take him from his primary task, research, and he was not prepared to forego that interest. This particular rationale is heard regularly in the halls of contemporary academe. At the university level, of course, teachers complain that their teaching load and committee responsibilities take time away from research. For some, of course, it is a specious complaint: they would prefer not to engage in research but are reticent to admit lack of interest. They would blame the system. Of course, at the level of the public schools, teachers rightly bemoan the lack of time to engage in private research pursuits. I think sometimes that action research was invented to provide opportunity to teachers to engage in scholarly activity even with their full teaching load. And though most states mandate continuing education, these obligations are too often viewed as yet another drain on the limited independence of the individual teacher, and these courses must too often receive minimum attention.

Spinoza, too, had to earn a living, and so he ground lenses, which must certainly have taken him from his philosophical research. And he repeatedly refused offers of money that would have relieved him of his pecuniary efforts; with these added resources he could have spent more time in his philosophical research. No, I do not think Spinoza refused the offer because teaching would have denied him time for philosophical research; rather, I think Spinoza believed that the offer of the professorship would have diverted his research from paths in which he would walk. He writes, "I am not holding back in the hope of getting something better, but through my love of quietness, which I think I can in some measure secure, if I keep away from lecturing in public" (375). There is, I think, a certain noise endemic to academia, and a certain quiet that comes from being away from its turmoils.

We are privileged at the university that we can shut our doors to achieve some semblance of peace; in the public schools, teachers retreat from the chaos of the classrooms, hallways, and lunchrooms to the lounges dedicated specifically to them. But these rooms, as teachers well know, are no respite from the school cacophony and provide little safe haven. The opportunity for study and reflective contemplation is minimal and not supported at these institutions. The demands and the geography of the school preclude the possibility of such effort. We are very busy, but too often, we do very little. William Pinar (2004, 252–253) cautions, "Too often we have mistaken busywork for academic work, authoritarianism for authority, indifference for professional dignity... We must abandon infantalized positions from which we demand to know 'what works.'" I think that Pinar's justified invective accurately depicts the environment Spinoza rightly rejected, one that reduces teachers to factory operatives, and that results in too many first-year teachers abandoning the profession, and too many seasoned professionals compromising their beliefs.

Thus, Spinoza's second reason for rejecting the offer seems akin to the first: he writes, "I do not know the limits, within which the freedom of my philosophical teaching would be confined, if I am to avoid all appearance of disturbing the publicly established religion" (252–253). That is, for Spinoza, being a teacher inherently situates the teacher in a paradoxical and inevitably impossible situation demanding the compromise of one's ideas and one's integrity. Yet, to be a teacher requires assuming a position often critical of the established cultural norm. Spinoza rightly understood that the state was not going to support the teachings of anyone who espoused the elimination of the state's authority and power, but that the proper role of the philosopher/scholar was to advocate for the assumption of the state's power by the state's citizens. Pinar (2004, 31) argues that

> Being a theorist does mean that the contemporary curriculum organization and the modes of cognition it requires must be bracketed, situated in history, politics, and our own life histories. Such understanding might allow us to participate in school reform in ways that do not hypostatize the present, but rather, allow our labor and understandings to function ... to enlarge the understanding and deepen the intelligence of the participants. The tragedy of the present is that school reform—as it is currently cast—cannot achieve this.

Indeed, it is to the maintenance of the present (what Pinar refers to as the "nightmare of the present") that the state desires its servants, of whom the teacher is one, to justify and support. Diane Ravitch (1995, 12) writes, "Content standards (or curriculum standards) describe what teachers are supposed to teach and students are expected to learn. They provide clear, specific descriptions of the skills and knowledge that should be taught to students." These standards, Ravitch (1994, xxvi) avers, must be based on "what is required for successful participation in higher education, the work force, and civil life." If the state sets the purposes of education, then the teacher in the classroom serves the state. This was completely antithetical to everything which Baruch de Spinoza believed in and about which he wrote. To the offer to assume the responsibilities of the classroom, Spinoza said, "I prefer not to."

There is another of whom I would speak who refused a public school teacher's life. Like Spinoza, in his writings and speeches, Henry David Thoreau meant to offer his fellow citizens a method for avoiding a life of quiet desperation. "Be rather the Mungo Park, the Lewis and Clarke and Frobisher, of your own streams and oceans; explore your own higher latitudes,—with shiploads of preserved meats to support you, if they be necessary ... be a Columbus to whole new continents and worlds within you, opening new channels, not of trade, but of thought" (438). It was learning that Thoreau desired, and it was teaching of which he spoke. In language remarkably similar to that of Spinoza, Thoreau wrote in *Walden* (1970) that he went to the woods to "live deliberately, to front only the essential facts of life, and see if I could not learn what it had to teach, and not, when I came to die, discover that I had not lived. I did not wish to live what was not life, living is so dear; nor did I wish to practice resignation, unless it was quite necessary" (222). And Thoreau, as did Spinoza and Marx and Dewey, saw the ideal philosopher as not one who only talked about the world, but as one who actually changed it. "There are," Thoreau accused, "professors of philosophy, but not philosophers...To be a philosopher is not merely to have subtle thoughts, nor even to found a school, but so to love wisdom as to live according to its dictates, a life of simplicity, independence, magnanimity, and trust. It is to solve some of the problems of life, not only theoretically, but practically." To be a philosopher, then, is to be a teacher. But to work in the schools, Thoreau discovered, is not to be permitted to teach.

Thoreau's learning directly concerned the fulfillment of living, a goal Dewey espoused as the ends of an education. Thoreau's ideal was a life fulfilled in a society that made possible a fulfilled life.

> I learned this at least, by my experiment; that if one advances confidently in the direction of his dreams, and endeavors to live the life which he has imagined, he will meet with a success unexpected in common hours. He will put some things behind, will pass an invisible boundary; new, universal, and more liberal laws will begin to establish themselves around and within him; or the old laws be expanded, and interpreted in his favor in a more liberal sense, and he will live with the license of a higher order of beings. In proportion as he simplifies his life, the laws of the universe will appear less complex, and solitude will not be solitude, nor poverty, poverty, nor weakness weakness. (*Walden*, 440)

I think Spinoza and Thoreau and Dewey would have been welcome companions.

Thoreau, we are told, quit his first teaching appointment because as a disciplinary measure he was required to cane his students. Edward Waldo Emerson (1999, 9) writes that one day Thoreau was being observed by one of the administrators of the public school in Concord, Massachusetts. We teachers know how these visits often go. "Deacon ____ sat through one session with increasing disapproval, waiting for corporal chastisement, the corner stone of a sound education, and properly reproved the teacher. The story which one of Thoreau's friends told me was, that with a queer humour,—he was very young,—he, to avoid taking the town's money, without giving the expected equivalent, in the afternoon punished six children, and that evening resigned the place where such methods were required." In a letter to Orestes Brownson, Thoreau writes, "I have even been disposed to regard the cowhide as a nonconductor. Methinks that, like the electric wire, not a single spark of truth is ever transmitted through its agency to the slumbering intellect it would address." With his brother,

John, he next opened a private school for the children of Concord, but his methods and philosophical position alarmed the children's parents. Indeed, it is said that Ellen Sewall, a participant in the school, and to whom both Thoreau brothers proposed marriage, was ordered by her father to refuse both offers. The school closed.

Teaching figured prominently in the lives of both Spinoza and Thoreau, but being a teacher was anathema to each. In "On the Improvement of Understanding," Spinoza (1955, 6) writes that "This, then, is the end for which I strive, to attain to such a character myself, and to endeavour that many should attain to it with me. In other words, it is part of my happiness to lend a helping hand, that many others may understand even as I do, that their understanding and desire may entirely agree with my own." Thoreau too wished to teach. In a letter to Orestes Brownson, Thoreau (1999, 29–30) writes, "I would make education a pleasant thing both to the teacher and the scholar...We should seek to be fellow students with the pupil, and should learn of, as well as with him, if we would be most helpful to him. But I am not blind to the difficulties of the case; it supposes a degree of freedom which rarely exists." It was this freedom that Spinoza *and* Thoreau sought in their philosophy so that learning could exist, and it is these satisfactions of teaching that Eisner proclaims. But our teachers prefer not to teach in the public schools; that environment is anathema to their efforts.

For the most part, Thoreau and Spinoza suggest, the school fosters not learning but miseducation. Thoreau writes, "How many schools I have thought of which I might go to but did not go to! Expecting foolishly that some greater advantages or schooling would come to me." And Thoreau's lineal descendent John Dewey (1964, 401–402) attributes the division of mind that current curriculum fosters to "stern discipline, i.e., external coercive pressure...motivation through rewards extraneous to the thing to be done...everything that makes schooling merely preparatory...ends being beyond the pupil's present grasp...exaggerated emphasis upon drill exercises deigned to produce skill in action, independent of any engagement of thought—exercises have no purpose but eh production of automatic skill...What do teachers imagine is happening to thought and emotion when the latter get no outlet in the things of immediate activity?" Dewey implicitly criticizes the teacher for this sorry state of affairs, but in our contemporary climate, in which the teacher serves only the state, it is the government which must bear responsibility. What both Spinoza and Thoreau said of the parochial schools is eminently true of the public schools today. In his *A Walk in Canada* Thoreau (2001, 285) writes that "We saw one schoolhouse on our walk and listened to the sounds which issued from it; but it appeared like a place where the process, not of enlightening, but of obfuscating the mind was going on, and the pupils received only so much light as could penetrate the shadow of the Catholic church." For Spinoza as well, little value could be derived from the school building, and for Spinoza, the Romish Church, was no less a barrier to freedom, than our present educational quagmire.

We have too much demand these days for professors of philosophy, and too little desire for philosophers. Our contemporary standards "bestride the narrow world / Like a Colossus, and we petty men / Walk under his huge legs and peep about / To find ourselves dishonourable graves" (*Julius Caesar*, I.ii.35–38). Pinar (2004, 30) accuses: "We teachers are conceived by others, by the expectations and fantasies of our students and by the demands of parent, administrators, policymakers, and politicians, to all of whom we are sometimes the 'other.'" Gerald Bracey notes that as a result of the strictures imposed by the legislation known as NCLB, "the beatings will continue until morale improves." Bracey (2003, 16) warns that the incessant testing will ensure "that a great deal of time will be spent preparing for the test and that a great deal of attention will be given to the results. Teachers will stifle thought, discussion, and question asking in the name and hope of raising test scores. Call it educational terrorism. I can't think of a better way to destroy a nation." Who would such fardels bear! Too many prefer not to.

Advocates of standards and measurements continue to insist not merely on the imposition of national standards, but on the imposition of *their* national standards. The ideological debate concerning the history curriculum points out the overtly ideological nature of curriculum. In her argument advocating the imposition of national standards in education, Diane Ravitch (1994, xxvii) urges that "the standards should be limited only to what students should know and be able to be well prepared for subsequent study, higher education, civil participation, and technical careers." Teachers have, thus, little freedom in the classroom, and no incentive to assume a freedom. The agglomeration of these autarkic achievements would lead, it is assumed, though not here stated, to a life fulfilled. Thoreau's complaint that education turns a meandering brook into a straight cut ditch depicts exactly the effects of such an education. Ravitch's list reminds me of the arguments of advocates for the subskill approach to reading: if each discrete skill of which reading is comprised is properly taught and learned, a student will become a good reader. Even E. D. Hirsh (2006, 11), though he mislabels this approach as romantic, decries this educational ideology. In his advocacy for a knowledge-based curriculum[9] he accuses, "I call this romantic idea,—'formalism'—a belief that reading comprehension can best be improved by acquiring formal comprehension strategies, not by building children's knowledge base." This belief holds the school in isolation from the whole of the student and teacher's life, and reduces learning to the acquisition of sterile and disconnected shards of memory.

More, the assumption that if only we would teach students would learn is wholly fallacious. On the one hand, it is based in a misapprehension concerning the process of learning. This is a very complex issue, but let it rest for now with Joseph Schwab (1971, 496): "The child with whom we work is both more and less than the percentile ranks, social class, and personality type into which she falls...we will...teach 'Tilda well only as we take account of conditions...which are not included in the theories which describe [her] as a learning child...We must take notice of these conditions, make some estimate of the relevance of each to the task in hand, and devise some means by which to cope with them." Even Terril Bell (in Bracey, 2003, 64), under whose tenure as secretary of education *A Nation at Risk* appeared with its vicious attack on the school system and its teachers, has written apologetically,

> We have placed too much confidence in school reforms that affected only six hours a day of a child's life...In the face of many negative influences on our children that come from outside the school, we have done well to maintain our high school completion rate and our level of performance on achievement measures...We have foolishly concluded that any problems with the levels of academic achievement have been caused by faulty schools staffed by inept teachers.

Nonetheless, the attacks continue without pause. We are a profession besieged and vilified. We teachers are continually accused of incompetency and blamed irresponsibly for the nation's problems. We teachers are overstressed and underpaid. We lack self-respect and are offered none by the public. As William Pinar (2004) says, "The present historical moment is, then, for public-school teachers and for those of us in the university who work with them, a nightmare" (3). We must struggle out way out.

His most Serene Highness the Elector Palatine, my most gracious master, commands me to write to you, who are, as yet, unknown to me, but most favorably regarded by his Most Serene Highness, and to inquire of you, whether you are willing to accept an ordinary professorship of philosophy in his illustrious university.

I prefer not to.

Chapter 3

On Finding Lost Objects

I am finishing my sixth decade. Every third thought is of the grade.

I find lately that I am starting to lose things. I grow conscious of a surprising number of single socks in my dresser drawers, and my kitchen shelves are filled with too many topless Tupperware containers. I cannot find my second favorite sweater, and there are computer files that are forever lost to cyberspace or on too many unlabeled mini disks. Books I stacked somewhere to read are no longer where I placed them, and books I had definitely read are missing from my bookshelves. I lent Phillip Roth's *My Life as a Man* to a dear friend to read. I regretfully remember that my dear friend has died. The book, too, alas, is gone. Not long ago I wrote a paper called "The Book That Changed My Life," about S. Y. Agnon's short story, "The Book That Was Lost." There I said that "I think that the book that changed my life is the book that was lost. I am certain that this book is a modest commentary on the complex codes (themselves...already a commentary on previous codes) that bound my life; this, alas now lost book would afford me precious insights into the complexities and questions that beset my daily existence. I read regularly to find this book." Sometimes, as Dylan said, "I can't remember the sound of my name." Lostness has become an important issue in my life as I enter this decade.

I have spoken earlier of beginnings and origins. Later, I will speak of endings and destinations. Here, I would think about the daily life in the classroom. I want to consider that what I have done in my life in schools is to pursue lost objects. Now, I do not hold with Plato that my quest for these objects is in the search for knowledge lost—I would not place so much burden on the infant, nor condemn the birth process to such trauma. Besides, we have learned that knowledge is not static nor monolithic; how can I lose what has not yet formed? Indeed, I have discovered that knowledge cannot be lost, it can only found. And I have come to believe that knowledge becomes so only when it *is* found and not a moment before. I would like to consider that all of our learning occurs in the pursuit of lost objects, and that this activity does not occur without succor, though sometimes it occurs in solitude. Knowledge becomes knowledge in, and as a result of relationship, and not in isolation. It is not truth really, but knowledge that will set me free, and my freedom happens with the aid of others, even as I am responsible for the freedom of others. Each day that I rise I do so with the intent to search for lost objects, and I wonder who will help me, and who will I assist in finding their own lost objects. Who will help me ease my burdens, and whose burdens might I relieve? Who will help free me from my human bondage; who will facilitate in my relief from captivity? Spinoza (1955, 248) says, "If we remove a disturbance of the spirit, or emotion, from the thought of an external cause, and unite it to other thoughts, then will the love or hatred towards that external cause, and also the vacillations of spirit which arise from these emotions, be destroyed." We find in learning what we did not know was lost, and we, then, make ourselves free.

I have spent most of my life in the classroom, and I have learned there the surprise of discovery, and I wonder how it is that I can find something that was not lost. Perhaps it is that learning is the pursuit of something that fulfills, something that answers an unasked question. There was a time when I did not know that what I did in the classroom was to pursue lost objects; I thought I was filling my mind with knowledge in these and those classrooms made readily available, though some of the others sitting next to me seemed to be learning different things, and even sometimes nothing at all! Then, I believed nothing was lost, only not yet visible. Then, knowledge was illimitable, and time was my only nemesis. Then, I had earned a few academic degrees, and I knew so many things. Ah, but I was so much older then.

I am still plagued by time, but it is not that against which I struggle. As my learning is endless, whatever I learn, in the words of the Haggadah, *dayenu*, it would have been enough. But what is it I do when I learn has struck me as an interesting question of late, and my study suggests, as I will discuss below, that when I had learned and as I learn now, I am pursuing lost objects, even if what I have found I did not know had been lost. We all have our pasts, and how we know and relate to those pasts very much depends on what we can find out about them. Learning about our selves and learning about the world is this process of discovery. Adam Phillips (1994, 69) says that "one of the aims of psychoanalysis [is] to produce a story of the past—a reconstructed life history—that makes the past available, as a resource to be thought about rather than a persecution to be endlessly re-enacted…[It] is worth wondering, at any given moment, what kind of object the past is for us…and so what kind of relationship we can have with it." As we discover the past, we find what we have lost, and reconstruct ourselves in this brave new world that has such people in it.

I think that we are always burdened by the past that we don't remember, and about which we possess no knowledge. Freud taught that in repetition we enact the past without remembering it; we are burdened and enslaved by it. Yes, those who do not remember the past are, indeed, condemned to repeat it, though for the most part they have no sense of the repetition. But education promises that folly is not in wisdom, but in ignorance. In the classroom we search together for the pasts we did not know were lost, and in the process, we remake ourselves and our world. Education promises not only that it is not too late to seek newer worlds, but that such pursuit has value.

Three Stories of Finding and of Losing

It was a beautiful wedding. The bride and groom were both terribly young, carefully dressed, well mannered, and ebullient. Because there were apparently few guests other than the couple's friends, and they were all equally young, I wondered perhaps the respective families did not approve of either the match or the event, or even both. Nevertheless, there was a great deal of excitement and, as at most weddings, I suppose, an air of anticipation and hope. My own daughter was one of the honored guests, pretty enough to be a bride herself.

Well, actually, she could just as easily have been getting married this evening as could any of her friends, because this was a mock-wedding, organized by the youth group of Beth Jacob Congregation to introduce the young teenagers to the solemn and joyous ritual rites of the wedding ceremony that joins a Jewish man and woman in holy matrimony. This was an event meant to discuss relationships in a relatively authentic setting. It was fun.

Of course, after the ceremony, a wedding reception was held, and I had volunteered to assist as wait-staff, cook, and bottle-washer. A friend of mine and I thawed the frozen hors d'oeuvres, poured the nonalcoholic drinks for appropriate toasts, and set out the cake to be cut. It was a lovely spread, and everyone wished the couple well. Guests danced to someone's iPod, and a good time was had by all, as they say.

During the clean up—the bottle-washing part—a smartly dressed, attractive woman approached me. I knew she was the parent of neither the bride or groom; perhaps she was the chauffeur who would whisk them away home. But it was me she approached.

"Alan," she said, "You probably don't remember me. I'm Sarah Naumburg. You were my tenth-grade English teacher! I was in your class."

This entrance from a long ago had never happened to me before, and so I was caught up short how to respond. No, I didn't remember her at all, but then, she was speaking of a time twenty-five years ago. I was so much older then! If for her I had not changed, she had for me dramatically altered. I could see neither her nor the classroom. But she remembered me her presence, the classroom's atmosphere, and even some of its substance. Though I could not remember her, she remembered for me the details. She remembered things I did not remember or know; she defined me for herself. She had found me, and she helped me find some lost objects. Did I need to remember these objects? Of course not, but that she helped me find these memories changed me irrevocably in ways I will never know. I once was lost, but now am found.

I received an e-mail. The subject line said, "Remember me?" This time I did recall, though I really cannot explain why this one and not the other. It was from a student long since lost to time and familiarity. She had Googled me (I do not know what hobbyhorse provoked this brazen act!), and discovered me here in Wisconsin at the University. Apparently, one is eminently findable in university settings. Unlike in the previous anecdote, as I have said, this student I remembered, though I cannot tell what hobby horse inspired this act of memory, and she wanted to tell me that I had been one of her favorite teachers, though, in general, she added, high school for her had been a horror. She said that she had been unhappy and afraid most of the time, and that she wanted me to know that my class was sometimes a safe haven for her. Of course, she admitted, much of the time that she sat in my classroom, she had been quite stoned, but nonetheless, she had learned a great deal from me, and wanted to thank me for my presence.

I hadn't a clue. Did I need to remember these objects? Of course not, but that she helped me find these memories changed me irrevocably in ways I will never know. William James (1962) writes that "things which we are quite unable definitely to recall have nevertheless impressed themselves, in some way, upon the structure of the mind. We are different for having once learned them" (69). I am enriched by this discovery. I once was lost, but now am found.

Before there were iPods (I really have lost memory of the time before iPods!), I was listening to NPR, and another of those talk shows concerning faith. I am interested in issues of faith: I have been running on faith for a number of years now. Given the political reality, what else can a poor boy do but live on faith when so much hope is gone? I have lost the sense of the program now, but at the end of the program, the anchor of the show announced himself and invited his listeners, if they had desire, to contact him at his e-mail address. The host's name was very familiar. If this was who I thought it was, then he had been a student of mine almost thirty years ago. Then, he had asked me to sponsor a school publication focused mainly on political issues. I agreed, and we gathered about us a small group of politically conscious people willing to write for and produce *North Star*, which was the name we gave to the journal. There was little budgeted money for the endeavor, and in those days when all the monies were earmarked for Air Force bombers and the schools had to sponsor bake sales, we embarked on a fund-raising campaign. For reasons long since lost, we decided to sell oranges to the community. We learned that we could buy the oranges at the Bronx Fruit and Vegetable Market at a dramatically reduced rate, and sell them at a reasonable profit. We took orders. And one early morning—it must have been about 3:00 am, students arrived at my apartment, and we six or seven late night souls piled into two cars and headed out on the Bronx Parkway en route to the Fruit Market, passing along the way a population whose lives were based in the streets and alleys of that market where most of New York's vegetables and fruits were sold while most of us remained asleep. I had no idea.

We purchased a great many crates of oranges without the slightest idea what we were actually doing, and returned to my home to put them in bags to be delivered later that day to our gracious customers. Oranges do not keep forever. I think we finished our work at about five or six o'clock in the morning, and they all went home to sleep. I showered and dressed for school. We earned enough money from the sale of oranges to publish yet another issue of *North Star*.

I had not remembered this adventure until I heard Brian's voice on the radio. Do I need to remember it? Of course not, but that Brian helped me find these memories changed me irrevocably in ways I will never know. I e-mailed him: "Was this the Brian Waldner who had been editor of the *North Star?*" He responded, "Yes, indeed, it is. Is this my eleventh-grade English teacher?" Indeed, I had been. I couldn't have been it without him. I once was lost, but now am found.

I have spent my life in the classroom; I wonder now what it is I might have lost and have been pursuing ever since.

In these nightmarish times in which everything has been found and must be delivered posthaste, I insist I seek what is lost, though here I acknowledge that too many times I have too easily found my way, and that I have too frequently lost my purpose, indeed. I have too often in the classrooms ceased looking for lost objects. In the routinization of my school life made too easy by the economic and political powers that have

assumed control over school policies, in the homogenization and corpora-
tization of academia, in my own self-absorptions and quietistic retreats, I
have, perhaps, lost the awareness of what I once intended to do there. In
the redundancies, formulaic prescriptions, and petty requirements of academic life, the excitement of discovery and contingency has been too easily
lost. It might have been that Thoreau (1970) "went to the woods to live
deliberately, to front only the essential facts of life, and see if [he] could not
learn what it had to teach, and not, when [he] came to die, discover that
[he] had not lived" (222), but he soon acknowledged that "It is remarkable
how easily and insensibly we fall into a particular route, and make a beaten
track for ourselves. I had not lived there a week before my feet wore a path
from my door to the pond-side; and though it is five or six years since I trod
it, it is still quite distinct. It is true, I fear, that others may have fallen into
it, and so helped to keep it open" (439). In the familiar and easily known,
there is nothing lost and no mystery; there the possibility for awe and wonder is diminished. In the absence of awe and wonder, the sense that there
is something to be found is lost. "I left Walden," Thoreau tells us, "because
I had other lives to lead." Familiarity breeds not contempt but disregard.
If I have lost my sense of wonder and awe, what I might have lost in the
classroom is a sense of purpose.

But on my better days, I know better. I know where I am. And I have
come 'round to think and to teach that what I am doing in the classroom is, indeed, pursuing lost objects. Long ago, I realize, I abandoned
the search for answers, and settled into the security of the question. The
question reminds me that I am seeking something, and if I now choose to
call that something "lost objects," well, I have my reasons. If, as I believe,
it is the question that opens up the conversation, then that conversation
requires the presence of others. Perhaps that I have lost something and
seek its recovery requires the presence of others, and the complicated conversation in which I engage is in the pursuit of those objects. I cannot
discover them on my own; what I seek in the classroom is not so much
community as communality. In some way that interests me to consider
now, in the endless pursuit of the question, I have been given help by
many others. I have been the help for many others in the endless pursuit
of their questions. And it strikes me now that what I have sought all along
in the classroom is the mutual relationship deriving from the pedagogical
interaction. Who cares what it is I have lost! It is the *process* of finding
that holds meaning. Whatever I find, it will suffice, and be what I was
looking for.

Henry David Thoreau spoke of losing things. Early in *Walden* (1970) he
writes, "I long ago lost a hound, a bay horse, and a turtle dove, and am still
on their trail" (157). What Thoreau might have meant by this statement

has excited readers since *Walden* was published. But that is, I think, the nature of the lost: it requires pursuit and it is personal. Finally, of course, what Thoreau has lost is an unanswerable question. Though many wonderful words have been written exploring what Thoreau might have meant, it is a fruitless search. As *Walden* is the record of a man advancing confidently in the direction of his dreams and endeavoring to live the life he has imagined (440), his pursuit of the hound, the bay horse, and the turtledove can only be defined as the activity of his life. These objects will never be found, though the search is unbounded. The hobbyhorses of Sterne's *Tristram Shandy* or the psychic genera of Christopher Bollas (1992) speak to the particular items each of us has lost and of which we live in pursuit. These hobbyhorses or psychic genera—or hounds, bay horses, or turtledoves—might be understood as the unconscious ego's structures based in a necessity for "inner organization, pattern, coherence, the basic need to discover identity in difference without which experience becomes chaos" (Milner, 1987, 84). These psychic genera are the nucleus by which the world might be organized, from which an outlook on life may be created, and from which new questions and works may be secured. These genera facilitate engagement with the world by promoting contact in it to evoke affective and ideational states that are the self. They are understood as bits of experience—ideas, words, images, experiences, affects—that evoke psychic interest, and that subsequently "scan the world of experience for phenomena related to such inner work" (Bollas, 1992). These genera send us out in search of what has been lost. When we find those objects, we will know it; though we know we are searching, we do not always know for what or why we search. That understanding has been lost. We search nevertheless for what can be found.

It might be discoverable *why* horses hold such sway over Alan Strang, in Shaffer's drama *Equus*, but there is no explanation why *horses* have ever attained this power. The psychiatrist, Martin Dysart (1973, 76) says, "Moments snap together like magnets forging a chain of shackles. Why? I can trace them. I can even with time, pull them apart again. Why at the start they were ever magnetized at all—just those particular moments of experience and no others—I don't know. *And nor does anyone else.* Yet *if* I don't know—if I can never know that—then what am I doing here? I don't mean clinically doing or socially doing—I mean fundamentally! These "whys" are fundamental…" That original connection, the energy that activates a hobbyhorse, that speck of dust that is the psychic genera and about which the pearl might be spun, is forever lost and may be never found; but we go on seeking, and the pursuit that the search instigates engages life. The motives for the quest, and the identities of the objects that are sought, are our lost objects, and the pursuit of either and both constitutes character. I would answer Dysart that the activity in which he is fundamentally engaged is helping Alan discover some lost objects, so that Alan might become character.

I have similar issues in the classroom. Why does *this* child learn so easily and *that* other child seems so incapable of learning at all? What lost objects is each seeking? Why does this child respond this way to me, and that child in another way? Why does this child seem so curious about this subject, and that child simply cannot be bothered? Why does that child read science fiction and that other child only science? I DON'T KNOW. *And nor does anyone else.* Yet, *if* I don't know—if I can never know that—then what am I doing here? I don't mean clinically or socially doing—I mean fundamentally!!

It might be said that Alan Strang's life is organized by that lost question: why *horses?* And my life, too, has been organized by that lost question: why *school?* I can explain ex post facto why I engage in learning, but I do not know why it is education that has satisfied my desires and organized my pursuits. It does, nonetheless. Alan Strang had too few to help him discover his lost objects, and too many who would maintain him in his captivity. He knew neither that he sought lost objects, nor even that there were objects that he had lost. Finally, through a series of horrific and self-destructive events, Alan is sent to Dr. Dysart, who cares enough to help him find his lost objects and redeem him from captivity. Interestingly, Dysart, too, pursues lost objects. I have been quite lucky in my search, and have had many who helped me find my lost objects and released me from my captivity. That is always how it must be: someone must help us in our searches.

Here, I would speak of the losing and the lost, of the burdened, of the captive, and the free. I have spent my life in the classroom; I wonder now what it is I might have lost and have been ever since pursuing. I know that I have had great support in that discovery.

I have spent my entire adult life in the classroom. Except for two horrific years in the factory, I have not left academia. I have spent my life in the classroom; I wonder now what it is I might have lost and have been pursuing ever since. Perhaps, it has always been another question. I can't remember.

In my family, we have been telling this joke with some success for years.

A very elderly couple is experiencing understandable problems with their memories, and so they decide to go to the doctor to ask his opinion. They describe for the doctor the problems they are having with their memory. The doctor gives both of them complete physical examinations, tells them that they are healthy, but that as one ages, memory tends to weaken. Perhaps, he suggests, they should begin to write down things to help them remember.

Later that night while watching TV, the old man rises up from his chair. "Where are you going?" his wife asks. "I'm going to get some ice cream." She requests a bowl for herself. "Sure," he responds. "Don't you think you should write it down?" she asks sweetly. "No," he answers, "I can remember a simple request like that." "Well," she adds musingly, "could you put strawberries on top of mine?" "Sure," he responds. "And please," she adds, "write it down so that you don't forget!" He says, with a tad of irritation in his voice, "Stop telling me that. I think I can remember that little thing." "Hmnn," she says, "well then, can you put some whipped cream and chocolate syrup on mine as well? And this time I know you should write *that* down because you will surely forget." With not a little irritation in his voice, her husband barks, "I can remember that, please." And the elderly man stomps off to the kitchen, from which a cacophonous noise of pots and pans emanates.

Twenty minutes later the old man returns to the television room and hands his wife a plate of bacon and eggs. She stares for a moment at the plate, looks up at him with some frustration, and says to him, "See, you forgot my toast."

I am writing things down.

The text about which I am writing is folios 33 (a,b) of Tractate *Bava Metzia* of the Babylonian Talmud, circa 200–600 CE. In *Bava Metzia,* the Rabbis concern themselves primarily with laws regarding claims arising out of any transaction in which two parties have a share. The particular *sugya* I mean to explore here deals with just such an issue: the prioritized obligations owed to an Other in helping him or her to find lost objects, to lay down his or her burdens, and to free him or her from captivity. The Rabbis wonder, given multiple demands for assistance in these endeavors, to whom we must assign first priority. And remarkably enough, in the Rabbis' discussion concerning these obligations, the character and role of the teacher plays a crucial role in the matter. Though *Bava Metzia* as a whole concerns issues of trade and industry, this particular sugya about finding lost objects, relieving burdens, and redeeming from captivity might not be so strange after all as a place to consider education and teachers: teachers *and students* complain regularly of their burdens, and students and teachers regularly describe their situation as being "in captivity." Finally, I would like to consider that what we do in school in the classroom is to forever pursue lost objects—our hounds, bay horses, and turtledoves—and that in schools we teach to others that this pursuit and effort is both a personal and a communal obligation.

Lost Articles

An earlier section of Talmud (30b) had suggested that by a lost article, the Rabbis mean an item whose purpose remains unrealized. "What is a lost property?" the Mishnah asked. But rather than provide definition, the Rabbis provide examples. Lost property is *not* an ass or a cow grazing by the side of the road, but only one whose trappings are overturned. A spade lying by a side wall is not lost, but a spade lying on a road might be considered so. A cow running among the vineyards is lost. "A cow grazing by the road for up to three whole days is not lost, but beyond that time it may be considered a lost article." A lost article, then, seems to be defined as an item whose purpose necessitates a relationship. The difference between a cow grazing by the side of the road—an animal that is not lost; and a cow grazing by the side of the road whose trappings are overturned—an animal that is certainly lost, is that there is nothing about the former that suggests its isolation, its lostness, whereas the overturned trappings give indication of a missing piece. Lostness is a function of separation from that which gives meaning. And what appears to give meaning here exists in a relationship to the Other. A lost cow is good for nothing. That is, the cow requires the Other for realization. The spade lying by the side of the wall has been left *by* someone; the presumption is that that someone will return. A lost article is unrealized—or better—unrealizable potential. Christopher Bollas (1992, 59) writes that "[T]o be a character is to enjoy the risk of being processed by the object—indeed, to seek objects, in part, in order to be metamorphosed, as one 'goes through' change by going through the processional moment provided by any object's integrity." In fact, all by myself, as a teacher, *I* do not exist: I am, indeed, lost. But once I am found and begin to exist, I, the once-lost article, am called into action in relationship. A relationship then, means responsibility. In discovering lost articles, we engage in relationships and become responsible.

I always require the other for identity. I am always lost and waiting to be found, and the discovery by the other ensures my existence. Dewey (1974, 430) long ago argued that "the school is primarily a social institution . . . [and] education being a social process, the school is simply that form of community life in which all those agencies are concentrated that will be most effective in bringing the child to share in the inherited resources of the race, and to use his own powers for social ends." Everything in school, then, might be considered as organized about the return of lost objects, and if schools have purpose, the finding of those lost objects is the primary goal. Education might be considered an engagement in the finding of lost articles. Books may be written, but if they are not read, the author is lost until he or she is found by a reader. Knowledge becomes so only when it is found by a student. Without students, teachers are, indeed, lost. The Rabbis wonder what obligations govern the finding of lost objects, and what do those obligations mean for the work in which we engage?

Plato and Socrates, the tradition in which I have been educated and in which I practice, suggest that, indeed, we have lost all of our knowledge, and we must all spend our lives retrieving it. Education here appears to be a matter of interest; it accrues to the individual engaged in it as a sort of quantitative measure. I believe the conservative right (an ironic term, but one that the politics of our times necessitates) would argue that the retrieval (or even acquisition) of this knowledge is in the interest of the individual and the State. Today's public discourse consistently links test scores to national standings in international relations. But for Socrates and for the present-day United States, education is not discussed as a transaction in which both parties share. That is, education is what students receive; knowledge is what teachers transmit. Knowledge is about ownership, and not about relationship. Interestingly, Socrates disowns his role as a teacher and disavows any share in the educational endeavor. For Socrates, knowledge may be a thing we have lost and must be recovered, but we must acquire that knowledge by ourselves alone. Socrates argues that if learning occurs, it is the responsibility of the learner to accomplish it; Socrates, for one, would not be held responsible. "If any one of these people becomes a good citizen or a bad one, I cannot fairly be held responsible, since I have never promised or imparted any teaching to anybody, and if anyone asserts that he has ever learned or heard from me privately anything which was not open to everyone else, you may be quite sure that he is not telling the truth." Two things seem evident here; first, for Socrates, truth seems to be available to the seeker as a product to be owned, so to speak; one has only to look properly for it. And second, whatever might be lost and/or gained by the other is of little interest to Socrates. In the educational transaction, Socrates disclaims his participation as either buyer or merchant. The Rabbis will insist, however, that *both* parties have a share in the activity under discussion. Thought about in these terms, education might consist of a personal obligation to an Other. Education might be understood as an ethical enterprise.

More directly, under the influence of this Talmudic discourse, I want to consider that education might be not an obligation *to* another—which suggests positions of inequality—but rather, an obligation *for* another. The relationships that education entail require that each party has an equal, though not identical share, and that the responsibility of each in such an encounter consists in being available for the other for the discovery of lost articles—wherever they might be; in the relief of burdens, whatever their weight; and in the redemption from captivity, regardless of the effort or even the cost. In these activities both parties have a share. As it is moral, education is ontological; my engagement assumes a responsible being. If it is only law that obliges our concern, then ethics is replaced by that law; and it is to the law and not the Other to whom we are obliged. But if we act not because it is the law and have been so commanded, but because we have commanded the other to command us to act, if we are there *for* and not merely *with* the Other, then we are engaged in ethical behavior. I am I who is responsible for you, to find you, as it were, and you are who makes me responsible. In this relationship, we share in the event, and we create each other. Education is finally an ethical enterprise.

The Mishnah[1] reads:

> If [a man's] own lost article and his Father's lost article [need attention], his own takes precedence. His own and his teacher's—his own takes precedence; His father's and his teacher's—his teacher's takes precedence, because his father brought him into this world, whereas His teacher, who instructed him in wisdom, brings him to the future world. But if his father is a sage, his father's takes precedence. If his father and his teacher were [each] carrying a burden, he must [first] assist his teacher to lay it down, and then assist his father. If his father and his teacher are in captivity, he must [first] redeem his teacher and then his father, but if his father is a sage, he must [first] redeem his father and then his teacher.

Since all law must find its source in Torah, the Rabbis will find justification for at least the directive for finding one's own lost article. Interestingly enough, the Rabbis do not seem concerned with the biblical sources for the latter two mandates: for relieving burdens and redeeming from captivity. Perhaps, the responsibility for these acts is assumed to be part of the foundational ethic: "You shall not subvert the rights of the stranger or the fatherless; you shall not take a widow's garment in pawn. Remember that you were a slave in Egypt and that the Lord your God redeemed you from there; therefore do I enjoin you to observe this commandment" (*Deuteronomy*, 24:17–18). Finally, it with priorities that the Rabbis seem most concerned: all obligations, the Rabbis suggest, must ultimately be fulfilled.

Finding Lost Articles

"If a man's own lost article and his father's lost article need attention, his own takes precedence. His own and his teacher's—his own takes precedence." The Rabbis do not seem concerned with either the nature or value of the lost article. As I have said, I assume that we are in the realm of ethical questions: what are the obligations we pledge to another, in this case, for the retrieval of lost objects? What is apparently assumed—what is apparently beyond question—is our obligation *for* the Other. The Rabbis do not ask *if* we should assist in finding the lost article; rather, they want to know, given multiple obligations, whom to assist first. Because we are bound to be responsible for the Other, because we are unquestionably and illimitably obliged for the Other, we must, it would seem, establish a priority for the satisfaction of our obligations or we might remain frozen in our responsibilities. And besides, there should be no poor among you. Therefore, the Rabbis establish a formula for the setting of priorities. I do not believe that it is with the nature of the articles that we are really concerned, but rather, with the finding of them. It is the process and not the product that is the focus of the discussion. The Rabbis seem to be concerned with a matter different than the protecting of economic interests; they are interested in the responsibilities we have for others. Since we are infinitely responsible, then the Rabbis' discussion here suggests that unless these separate claims upon us are regulated, the legitimate strength of each responsibility might immobilize us from all action.

The Rabbis are defining ethical issues, and one of their concerns here, as I will show, is education. Why discuss education in the context of finding lost articles, of relieving burdens and redemption from captivity? Though the entire tractate *Bava Metzia* is concerned with trade and industry, this Mishnah refers to fathers and to teachers rather than to business partners. On the one hand, the Rabbis suggest that the same principles that apply to trade and industry apply equally to familial and educative relations. On the other hand, it must also be true that familial and educative relations be applicable to trade and industry. Perhaps this Mishnah does not distinguish a behavior particular to the workplace, home, or school: they are all alike. This Mishnah establishes principles upon which priorities of obligation might be adduced everywhere without qualification.

Interestingly, the Mishnah suggests that in this endeavor the first obligation is to self. Self is the actualization of potential. And so the Mishnah insists that we discover our own lost article first. That is, until we find our own lost articles and establish our relationships to self, until we activate our own potential, we ought not to undertake assisting others. This may be the reason the Mishnah seems to be redundant: it does not refer to a man's article but to a man's *own* article. The use of the word *own* suggests that the article is not a mere possession but an intrinsic part of the man. The lost article is defining. Perhaps the Mishnah suggests that we cannot find anyone else's lost articles until we find our own. And since a lost article is one whose potential is yet unrealized, finding our lost articles is the establishment of self in the world of objects. This appears to me to be the

Rabbis' early statement of what will become known as "object relations theory."

I find the Rabbis' establishment an interesting hierarchy—to be obliged to discover one's own article before that of either father or teacher—given the importance in Judaism of the fifth commandment, "Honor thy father and thy mother." Here the halakah[2] states that first you discover your own lost article prior to that of your father or your teacher. The Mishnah clearly affirms the priority of the needs of the self. So it must be that attending to one's own lost articles before attending to those of either your father's or your teacher's does not violate the command to honor the parents or the teacher. Indeed, one can only honor the father and the mother if the self is first attended to. Similarly, the Mishnah insists that our obligations to the teacher may not inspire us to action until we have first satisfied ourselves. The point seems clear: we must first be a Self to be responsible. To satisfy ourselves, we must retrieve our own lost article. The Mishnah implies that one cannot discover another's lost article while distracted with the self's own lost article; the Rabbis prioritize the needs of the self. The self is here actualized in discovering its lost articles. The self is realized in relationship to objects. What is lost, I recall, is always that whose potential is yet to be realized.

The Teacher

The Mishnah goes on: "His father's and his teacher's [lost article]—his teacher's takes precedence, because his father brought him into this world, whereas his teacher, who instructed him in wisdom, brings him to the future world." That is, our first obligation is to the teacher, to the future. The Rabbis describe the father's efforts as having been completed in the past, but the teacher's work is to be realized in the future: the father is responsible for the product—bringing one to this world, but it is the teacher who brings one to the future world.[3] The teacher, actually, could not offer a product because the future is intangible; like Godot the future will always arrive tomorrow. In this sense, the future might be prepared for, but may never be known. And when the future arrives, it is no longer the future. The Mishnah associates the future with the teacher because the teacher offers not the future as product, but as process: the teacher offers the means to attain the future. The teacher offers process because he or she cannot possibly offer what does not exist except in potential. However, it is clear that when the student helps the teacher find the lost article, both share in the enterprise—both attain the future.

In a sense, the future might be likened to a lost article—unrealized potential. Thus, the student attains the future by first helping the teacher find the teacher's own future. When the teacher's article is lost—his own means—it is essential that the student help in its discovery; then the teacher might offer the student the means to the future. But perhaps even more, the Rabbis suggest that the teacher requires the student to facilitate the future—without the student the teacher's future remains lost. Education here demands a teacher and a student; education is a relationship.

It appears that the Mishnah is not about lost articles at all, but about relationships: and the principles that order the finding and retrieval of lost articles establish the relationships between teacher and student and between past and future. To give priority to the father's lost article is to reach back into the past; but the Mishnah suggests that the self's first responsibility is to the future. Therefore, the relationship to which priority must be given is to that with the teacher. The past is over; my father remains my father regardless of the future. However, in establishing relations with the teacher, I prepare not only for the future, but for the present as well—in being a student, I create the teacher. Both have a share in the relationship. The Rabbis also know that until we have found our own lost articles, we have no self with which to establish relationships with the past or the future.

The Mishnah, however, qualifies these prescriptions. It states, "But if his father is a sage [or equal in wisdom to a sage] his father's [lost article] takes precedence." Well, now, it is clear that what distinguishes the parent from the teacher is the wisdom (we must deal with wisdom below—the Rabbis have something to say about that as well!) that is inherent in the teacher and that can be offered to the student. The priority here is always toward the future, but the past may not be discounted. Indeed, when the past might be linked with the future, it is that relationship that must be given priority.

Carrying Burdens

"If his father and his teacher were [each] carrying a burden, he must [first] assist the teacher to lay it down, and then assist his father." Perhaps the Rabbis acknowledge that once the lost article is found, it must yet be borne and ultimately, set down. Why are lost articles burdensome? Perhaps because a lost article as unrealizable potential weighs heavily. It is not enough to locate the lost article; it is not even enough to return it to its rightful owner. The burden of the article must be relieved; that effort requires a relationship. Or rather, the Rabbis prefer that a relationship be engaged; we know that most burdens might be simply dropped, but the relationship maintains the integrity of the objects that are burdensome and yet relieves the necessity of carrying them alone. It is the relationships upon which the Rabbis have focused and not the burden. The Rabbis insist that at least two people handle the burden. They insist that whatever the burden be, it requires assistance in its relief.

The Rabbis are no longer referring to lost articles; they have shifted their attention to burdensome ones. Perhaps the Mishnah make this distinction between lost articles and burdensome ones to ensure that it is not enough to find lost articles; our obligation extends to the relief of the other in carrying not only it but any oppressive object. As we all lose objects, so, too, do we all bear burdens. We have all left Eden where every thing was perfect and no effort was required. Here, the Rabbis seem to acknowledge that we live in this world and must bear our burdens of one kind or another here in this world. The Rabbis suggest that we all share responsibility for the relief of each others' burdens, and that everyone benefits from the engagement in the relationship.

Again the Mishnah affirms the importance of the teacher; for even if the father is also burdened, it is to the teacher's needs that priority is given. Of course, we are obliged to the father as well; responsibility is not exclusively given to the relief of the teacher's burden, rather, the Rabbis merely establish priorities for the relief of those separate burdens. Again, we are concerned with the commitment toward the future that exists in the relationship with the teacher. If we lay the burden down, we have relieved the teacher and together we might engage in more pleasurable activity. Or perhaps the Mishnah points out here that those who may give us the future must first be relieved of their own burdens. They can only achieve this freedom with the help of the student. The teacher requires the student's assistance in order to continue. We know this because the Mishnah specifically states that it is the teacher's burden that must be relieved, and, as we will see, only the student would recognize the teacher.

Of course, the Mishnah does not suggest here that the father's needs be ignored; rather, after the student helps the teacher, the father must be helped to lay down his burden. The past has its obligations; but it is the future that establishes priority. It is also interesting that the Mishnah does not suggest that we help carry the burden, but rather that we help in putting it down. Perhaps this is an acknowledgment that simply wearing another's burden does not bring relief for either party; it merely relieves briefly the weight. I am reminded here of a song by The Band entitled "The Weight": "You took the load right off Fanny and you put the load right on me." The Mishnah does not, like the song, advocate the transfer of burdens but rather, the relief of them.

Captivity

Finally, the Mishnah writes, "If his father and his teacher are in captivity, he must [first] redeem his teacher and then his father, but if his father is a sage, he must [first] redeem his father and then his teacher." Again, we note that both are to be redeemed, but the teacher must be redeemed prior to the father. Why is this? Perhaps because the teacher freed yet offers the paths to the future, but the father whether in captivity or whether free, still links us to the past. It is only when the past and the future and directly linked—as when the father is also a sage—that the father is to be redeemed before the teacher.

But why the progression from lostness to captivity and why don't the Rabbis address the relieving of the burdens of self or the freeing from captivity first of the self? Perhaps the Rabbis acknowledge that one's burdens cannot be relieved without assistance; nor is it possible to free oneself from captivity. Both enterprises require relationships. And I am intrigued as well by the Rabbis insistence that we redeem our teacher from captivity. Is it possible that the Rabbis recognize that an engagement to knowledge such as that held by a teacher could be captivating and lead her to a form of bondage. The Rabbis say that the student must redeem the teacher from this captivity—must draw him from the esotericism of his studies—and return him to the daily world. If the future is to be realized, then the student must redeem the teacher so that the teacher might offer the future to the student. Further, the Rabbis suggest that the future is engaged in the practical world, and not in the solitariness of study.

And so again the Mishnah affirms the importance of wisdom that is not inherent within but exists between. This Mishnah seems to suggest that education is a relationship that brings one to the future; mere familial obligation alone seems to hold one to the past. Yet Judaism has forever been called a patriarchal religion; we only recently have added to the patriarchs Abraham, Isaac, and Jacob the matriarchs Sarah, Rebekah, Rachel, and Leah. Yet in the Mishnah it is the relationship between a student and his teacher that takes precedence in the obligation. Within this Mishnah is a rent in the fabric that leaves space for tearing down the patriarchal order; for though it might be assumed that the teacher is male here, there is nothing in this Mishnah to preclude that teacher being a female. In that case the female takes precedence over the male: Sarah over Abraham,

Rebekah over Isaac, and Rachel and Leah over Jacob. And again we must understand that education is premised here as a process—it exists in relationships that lead toward the future.

Thus is the Mishnah, the law. But what authority undergirds this law is a concern to the doctors of the Talmud. All authority, of course, must reside in Torah, and be based in the overriding concern of *Leviticus* 19:2, "I am holy therefore you should be holy." The Rabbis now discuss, as is their wont those aspects of the law that interest them. I am interested in their interests.

If I Am Not for Myself

The Mishnah states the law, and I have explored the Rabbis' rationale for that law. But it is not enough to have a law: the Rabbis want to know upon what authority the law rests. Of course, final authority exists always in text; for the Rabbis these are the canonical texts of the Hebrew Bible. The Gemara,[4] here without attribution to any specific Rabbi, begins, "Whence do we know this?" I love this question: the Rabbis wonder, where do we get these ideas? It reminds me of my grandparents who would listen to me attentively and then say in Yiddish, *Vas es dos?* [Where did this idea come from?]. I think the Rabbis ask a legitimate question: you see, they seek human sources for human problems. It is not enough that some Rabbis— even Judah Ha-nasi—have established the priorities; these Rabbis seek textual authority. Thus, Rab Judah justifies looking after one's own lost article first by quoting from the Torah; he says, actually citing another Rabbi: "Scripture saith, *save that there shall no poor among you*: yours takes precedence over all others" (italics in original). Rab Judah has quoted from Deuteronomy 15:4. Rab Judah interprets the text to mean that there must be no needy person within the community; therefore, to neglect your own lost article ensures that there will be at least one needy person—yourself. Rab Judah concludes from this that one must find one's own lost article before attending to that of anyone else.

Interestingly, however, the Scriptural verse cited by Rab Judah refers not to the individual specifically, but to the entire community. The verse makes clear that there *will not be* a poor person in the Hebrew community "because Adonai will bless you in the land that Adonai your God is giving you as an inheritance, to possess." The complete passage in the Bible reads: "The foreigner you may oppress, and he who belongs to you; as for your brother, your hand is to release (him). *However, there will not be among you any needy-person, for Adonai will bless, yes, bless you in the land . . ."* (italics in original). That is, what in the Torah appears to be only a promise has been transformed in the Gemara by the Rabbis into a command. It would appear that here, in the Gemara, the Rabbis mean to insist that the self may not be ignored even for the good of the community. The Rabbis assert that, indeed, for the good of the community, the self must be looked to first. The Rabbis offer no advocacy of martyrdom; the concerns of self—in this case, for the discovery of lost articles—must be first attended to. Within the particular argument regarding lost articles, the Rabbis address a relationship between the needs of the individual and the needs of the community. The community's strength and well-being are only as strong as, and are dependent upon, the strength and well-being of the individuals who comprise it.

However, the Rabbis also acknowledge that without the community there are no individuals. Rab Judah says that if we do not look after our own lost article first, then there will be a needy person in the community—the self—and the community is seriously compromised. I think the Rabbis wisely acknowledge the unending obligation that the community might make on us—the illimitable lost articles that exist and that must be found. If we were to attend to them all, say the Rabbis, we would impoverish ourselves; but the Torah says that the community must not allow a needy person to exist within it. Perhaps the Rabbis recognize that every act contains self in it; even the apparent selflessness of the martyr is actually a selfishness; when we act for the community, we must not discount our own needs.

What Am I?

Rab Judah continues; "He who strictly [observes] this will eventually be brought to it." Rab says that if we only care for ourselves and do not think of community, we will eventually become impoverished. Obviously, then, poverty seems more than the absence of material means: here poverty is clearly the absence of relationship. If we are not obliged to helping others find their lost articles, we are poor as a result. Without the obligations to the other, we are individually impoverished because those obligations connect us to, and engage us with, the future. Obligation and Responsibility to the other do not precede self; they are themselves ontological. Or, as Zygmunt Bauman (1993, 71) says, "for is before with." My responsibility to the Other constitutes the Face of the Other. Without the other, my self is impoverished. What could that mean, an impoverished self? Aren't selves nevertheless selves, impoverished though they be? What could impoverished mean here, in this section on trade and lost articles?

I am reminded of Hillel's remarkable formulation of this principle in the *Pirke Avot*: "If I am not for myself, who will be for me? And being for my own self, what am I? And if not now, when?" The Gemara suggests that the finding of lost articles is a matter of relationships: for such an event to take place there must be one who has lost and one who can help find. But Hillel's famous question does not ask "Who am I?"; it asks "What am I?" Here the Rabbis suggest that poverty here is not a material or pecuniary one but, rather, situated in human essence. Outside of relationships we are not human—our obligation to the other makes us human. Levinas (in Bauman, 1993, 77) writes, "To be human means to live as if one were not a being among beings...It is I who support the Other, and am responsible for him...My responsibility is untransferable, no one could replace me. In fact, it is a matter of saying the very identity of the human I [starts] from responsibility." It is only in my responsibility for the other—my teacher—that I become unique. My responsibility toward my teacher makes me a student and propels me toward the future. I must recognize myself as responsible; I am human in my responsibility. Education then, in being ethical, is also ontological.

The Rabbis continue their discussion. They reintroduce the law: "If his father and his teacher were [each] carrying a burden, etc...." Interestingly, the Rabbis then engage not in a discussion of the nature of the burden, but with the definition of the teacher. They seem to assume we would recognize our fathers from our teachers, but how would we recognize a teacher? We all, it would seem, carry burdens; thus, if two people are carrying a burden how would we discern the one who is the teacher that we might first assist her? Or perhaps the Rabbis are interrogating who deserves to be called teacher. That is, what do teachers do that identify them as teachers? "Our Rabbis taught: The teacher referred to is he who instructed him in wisdom, not he who taught him Bible and Mishnah: this is R. Meir's view." Meir's view is that the criterion by which a teacher may be known exists in the capacity of him/her to instruct in wisdom. Clearly, however, that wisdom is not contained in texts—either the Bible (the ultimate text) or the Mishnah, the compilation of law. Rather, wisdom must be the insight with which those texts—or any text, for that matter—may be read. Wisdom is the capacity to discover meaning as a result of reading; wisdom resides not in a choice of texts, but rather, appears as a matter of intertextuality; he who can instruct in this practice may be adjudged as the teacher. This seems to agree with the sense of the passages in the Mishnah that mandate the retrieval of lost articles, the relief of burdens, and the freeing from captivity. Wisdom is the ability to connect the past with the future. Wisdom is to know the whole outside of the parts.

As I have come to expect, not all the Rabbis concur with this viewpoint. For example, speaking of the definition of the teacher, "R. Judah said: he from whom one has derived the greater part of his knowledge might be counted as the teacher." Interestingly enough, in R. Judah's view, knowledge has replaced wisdom, and knowledge may be quantitatively measured. R. Jose offers another opinion: "Even if he enlightened his eyes in a single Mishnah only, he is his teacher. Said Rava: like R. Zehora, who told me the meaning of *zohama listron*."[5] That is, Rab Jose ascribes to the teacher the ability to teach completely even *one* single Mishnah. Here, *engagement* is the criteria to measure a teacher, and its measure is a single learning, not a quantitative measure but a qualitative one that defines the teacher *and* the student.

Perhaps these are not as varied perspectives as it might at first seem. Perhaps the last—specificity of a single Mishnah—makes the second—knowledge of a text—possible; and the second is a condition for the first—attainment of wisdom, the insight into textualities. These may not be separate viewpoints at all, but different perspectives on the same viewpoint. Regardless of the locus of wisdom, it is the capacity to engage in a relationship such that communication and learning are enabled that designates the teacher. It is not *what* one knows that constitutes the teacher, but what one is capable of communicating. The teacher must exist in a relationship that is productive. This may certainly be opposed to say, Rorty's liberal education and the edifying discourses he advocates. Rorty says an edifying education should hold edifying conversations with as many different kinds of "hero-worshipers" as possible. If edifying discourse is that which permits us to talk about ourselves in different ways—Talmud isn't edifying; rather, it is ethical. The Rabbis suggest that wisdom should be the result of a relationship with the teacher, and not merely an acquisition of a new vocabulary to address cultural issues. Education is ethical and does not lead to ethics.

The Rabbis suggest that the teacher might be known in a number of ways—known not by her or his position in society, but by her relationship to the student. Indeed, it is the student, the Rabbis suggest, who will recognize the teacher. An interesting story is told. "Samuel rent his garment for one of the Rabbis who taught him the meaning of 'One was thrust into the duct as far as the arm pit, and another [key] opened [the door] directly.'" Textual commentary notes that this describes the way that the priests opened the gates to the Temple each morning: Two locks had to be opened. The first could be opened only by inserting one's entire arm through a hole by the side of the gate and opening the lock from the inside, while the second opened normally. Yet Samuel mourns this teacher's death for the wisdom contained in what seems to be this seemingly cursory teaching. What Samuel has learned hardly seems like wisdom; indeed, the Temple had ceased to exist in 70 CE, and the knowledge of how the priests opened the gates in the morning of relative insignificance. Why is this story here? Interestingly the Rabbis do not tell us why this was so important to Samuel. And perhaps this is not what is important here; rather what seems to be paramount is that Samuel recognizes his teacher. If we cannot define our teacher, then that person cannot be called one. The teacher is one who exists in relationship. And if relationships are ethical, if ethics is ontological, then education is the substance of being. For how many in my classrooms, am I not a teacher?

The next passage seems to accord with this position. Ulla says "That the scholars in Babylon arise before [before what I am not sure, perhaps prayer] and rend their garment for each other [in mourning]." Ulla acknowledges that students owe to each other respect for being each other's teacher. In exile, they are in mourning but, nonetheless, they behave as if they were in the presence of the teacher. That would suggest that they are responsible for the relief of each other's burdens. But then Ulla says that with regard to a father's or a colleague's lost article, one must return that of the teacher's first only if that teacher is one par excellence. I think that Ulla is establishing a hierarchy of teachers as well as acknowledging the importance of the father—the past—on the educative process. Ulla acknowledges that the students treat each other as teachers; but it is only the teacher who has instructed them in wisdom, the teacher par excellence to whom first priority is due. The distinction is important: engaged in education our colleagues are all our teachers, and we may be distracted by our view of the future and ignore the holds of the past; similarly, the hold of the past is strong and it requires a powerful presence to move beyond it. Only the teacher par excellence deserves priority because it is only she who can move us beyond the past; it is only *that* teacher whose priority supersedes that of the father. The Rabbis recognize how captivating the future seems and how easy it might be to be distracted from our responsibilities to the past. Thus they insist that it is only the teacher par excellence to whom first obligation is due.

But the Gemara complicates this discussion with a story about a disagreement between Rabbi Hisda and Rabbi Huna. Ulla has said that it is only the teacher par excellence who must be accorded priority over the Father. Hisda, however, suggests that a teacher par excellence is defined not only by the student's need of the teacher, but by the teacher's need of the student. Rabbi Hisda suggests that a teacher may be known not by what he offers, but by what he shares. Rabbi Hisda asked Rabbi Huna: "What of a disciple whom his teacher needs?" Hisda had been Huna's disciple, and by this question Hisda has suggested that his teacher had need of him. But Huna denies the obligation: "Hisda, Hisda, I do not need you, but you need me." And for the next forty years, these two Rabbis bore resentment and did not visit each other. The contiguity of these two passages suggests this: a teacher par excellence is one who acknowledges the obligation the teacher has to the student. To move into the future requires the other—and therefore every teacher par excellence must first be a student par excellence. Again we are led to the idea of education as ontological.

The Rabbis continue to disagree over the identity of the teacher: R. Judah opines that the teacher is he from whom one has received the greater part of one's wisdom, but Rabbi Jose has held that the teacher is *anyone* from whom one has learned even a *single* Mishnah. Certainly, Jose's opinion would expand the potential of many to be teachers, and therefore, offer an expanded ethical arena for action. One Rabbi argues that the weight of opinion rests with that of Rabbi Jose, though others object. Surely, the Rabbis say, Rabbi Johanan said, "The Halakhah is in accordance with an anonymous Mishnah, and we have learned 'His teacher who taught him wisdom.'" Of course, the citation of Rabbi Johanan might not necessarily contradict Jose's position: if the single Mishnah to which Rabbi Jose refers as the mark of the teacher is the one identical to that to which Rabbi Johanan refers, then wisdom might, indeed, be understood as the knowledge of a single Mishnah—the one that defines the teacher as one who instructed him in wisdom.

Hence, we have returned to the question concerning the meaning of "wisdom." Johanan clarifies: he states that the teacher is not he or she who offers "wisdom," but rather, he or she who offers "most of her wisdom." On the one hand, the teacher remains cognizant of the student, and offers that which the student is capable of learning: not all of the teacher's wisdom, but as much as the student can handle. On the other hand, perhaps Johanan suggests that no teacher can transmit all her wisdom because no teacher possesses complete wisdom. What I as a teacher transmit is also mine to have; therefore, if I, as a teacher, do not continue to learn then I cease to be a teacher—I am impoverished. If all I do is teach and not learn, then I abandon wisdom and I become the needy person in the community. Therefore the teacher is not only the one who instructs in wisdom, but also one who has wisdom. This is consistent with the Mishnah regarding lost articles. It is also a commentary upon the story of Huna and Hisda. Finally, it returns us to the idea that education is ontological: I am my learning and my willingness to learn.

And what is meant by the greater part of one's knowledge? Clearly it cannot be contained in a single text. "They who occupy themselves with the

Bible [alone] are but of indifferent merit; with Mishnah, are indeed meritorious; with Gemara there can be nothing more meritorious; yet always run to the Mishnah more than to the Gemara." As it is rightly pointed out, this appears to be a contradiction. If the greatest part of knowledge is in the Gemara, then why do the Rabbis instruct us to run more to the Mishnah than the Gemara? Rabbi Johanan responds historically: when the Gemara was touted as the most important text, then the Mishnah was abandoned; thus the Rabbi countered this teaching by demanding that the Mishnah be studied. There is more than psychology here. Rabbi is suggesting that without Mishnah—the text and the law—then commentary is meaningless, mere verbal pyrotechnics. Unless interpretation is attached to actual practice—as mandated in Mishnah—the interpretation is meaningless. Rabbi insists on the importance of interpretation, but insists as well that at the base of interpretation must be the text. We are once again joining past and future, even as we did above when we wondered about lost articles and burdens and captivities. The Rabbis know that the past has greatest meaning—perhaps its only meaning—when it may be joined to the future that does not yet exist but may be realized. The future, like the lost article, is unrealized potential. Wisdom is the joining of past and present. The teacher is the one who is capable of transmitting wisdom by being in relationship with the student.

Again the Rabbis ask, "How was that inferred?" That is, how is it we know that the study of Talmud—the law and the commentary—is the most meritorious effort? And Rabbi Judah turns remarkably to *Isaiah* 58 for his justification. There, God orders Isaiah to call the people and the house of Jacob to account for their transgressions and sins. There are two elements that the Rabbis focus on here: the distinction between *transgression* and *sin*, terms we might consider synonymous, and the distinction between *my people* and *the House of Jacob*, again terms that ought to refer to the same population. In the quoted passage, Isaiah is called upon to demand of the people vigilance even though the people already "ask counsel of me day by day, and say they delight in knowing my ways, although [they are] like nations which have acted rightly and not forsaken the just laws of their gods, they ask me for righteous laws and say they delight in approaching God." Indeed, Isaiah seems to be describing a devout people; yet God insists that the prophet demand from them continued vigilance. Lack of knowledge is not considered sufficient excuse for an act of sinning. Rab Judah asks, "What is the meaning of *Shew my people their transgression, and the house of Jacob their sins?*" And he responds, " 'Shew my people their transgression' refers to scholars, whose unwitting errors are accounted as intentional faults; *and the house of Israel their sins*—to the ignorant, whose intentional sins are accounted to them as unwitting errors." The transgressions of scholars lie in their beliefs that their answers are complete and final; they negate who they are—scholars—in the denial of the need for more study. Thus, their transgressions are accounted as committed willingly even though they are engaged in scholarly work. Sins are committed unwittingly and ascribed to the general populace, but are nevertheless attributable to lack of study and education. And those who in their ignorance of the law transgress willingly, they are not excused for their failure to know. The ignorant are yet culpable; they must learn and for this require teachers. Judah holds all to account for their behavior.

Interestingly, all transgression and sin are attributed, however, to lack of education; the errors of scholars because they did not study enough, and the errors of the general populace because they did not know enough to study. In this passage Rab Judah, son of R. Ila'i, distinguishes between "my people" and "the house of Jacob," between teachers versed in knowledge and the general populace. The Rabbis make this distinction between my people—those who are consciously committed to the study, which we have seen is ethical and ontological and whose purpose is to make a holy people—and the house of Jacob, the population that aspires to holiness, but that does not consciously study to be such. Continued study is required of all. In this interpretation, the Rabbis have placed education as the central determinant of holiness. It is education that will produce a holy people. Thus R. Judah says, "Be heedful of the [Talmud], for an error in Talmud is accounted as intentional." There is no end to study. It is inferred then that running to the Mishnah more than the Gemara is the most meritorious study because it is only when the two are conjoined that wisdom might be produced. But why would running to the Mishnah more than the Gemara be the most meritorious study? How might this be inferred?

In this passage from Isaiah cited by Rab Judah, God warns the people against the hypocrisy of obeying the letter and not the spirit of the laws, attending to the Mishnah without the Gemara—the isolation of their behavior in study from engagement in the daily context of their social lives. This behavior might lead to a form of quietism, and the Rabbis seem to acknowledge this possibility in the distinction between sin and transgression. The Rabbis suggest that a scholar must do more than merely study and acquire knowledge—he or she must *continually* study, and in that act acknowledge the incompleteness of knowledge. Finally, it is study linked to action—to the demands of social justice that is paramount. The rest is transgression. To declare the conversation ended and study complete, to arrive at finality is a transgression. The possibility for further study must always be acknowledged because the complexities of human existence are such that definitive answers are impossible. That is, the purpose of education must improve the quality of living for all, and that study which does not have this as its end cannot be deemed education; those who are not so engaged transgress and sin. This may be inferred because in this same chapter of Isaiah, God distinguishes between authentic practice and appearance:

> Since you serve your own interest only on your fast-day and make all your men work the harder, since your fasting leads only to wrangling and strife and dealing vicious blows with the fist, on such a day you are keeping no fast that will carry your cry to heaven. Is not this what I require of you as a fast: to loose the fetters of injustice, to untie the knots of the yoke, to snap every yoke and set free those who have been crushed? Is it not sharing your food with the hungry, taking the homeless poor into your house, clothing the naked when you meet them and never evading a duty to your kinfolk?

In the absence of action, study is all form; study must lead to practice.

The Rabbis have been talking in this piece of Talmud about responsibilities for the other, about teachers, and about wisdom: wisdom has been called the greater part of knowledge. We see now that this entails the acknowledgment of not knowing enough and therefore, of not doing enough. The Rabbis say here that the greater part of knowledge consists in the willingness to admit not knowing, of acknowledging not only that we do not know enough, but that we have not yet done enough. We must always continue our study because though we do not finally know, we must never cease to act even if we have not finished studying. We continue to study. We act. The Rabbis of the Talmud suggest that exegesis of text is unending—there is always another meaning to be explored as long as the Mishnah is considered. Thus, those scholars who commit unwitting errors do so *intentionally* because they have not sufficiently studied Talmud. That is why the Rabbis turn to Isaiah for their proof of what is a teacher and therefore, what is wisdom. This passage addresses hypocrites: the Rabbis apply it to scholars who serve their own interests and who are, therefore, impoverished—they deny community and therefore are not wise though they lay claim to the title.

The Rabbis seem clearly to suggest to me that continued and continual study of the Talmud—which is to say the creation of Talmudic discussion—is of greatest worth. That study links past and future; the future as a time in which social justice prevails. Education is grasped as ethical. Scholars—those engaged in this enterprise—must have faith that even in their ignorance, in the words of Isaiah, "we shall see you rejoice." The study of Talmud will raise the Hebrew people to be a holy people—and all those who have forsworn this study shall be ashamed. Judah asks, "What is meant by the verse, *Hear the word of the Lord, ye that tremble at his word.*" We might understand tremble not as the response to fear but to awe. Abraham Heschel (1962, 53) says that "The beginning of awe is wonder, and the beginning of wisdom is awe." Judah quotes from this verse in Isaiah to give strength to those who continue to struggle with learning and education despite the contempt and calumny of those who decry their efforts. Judah means to give comfort to scholars; it is they, as Heschel says, who recognize in their awe at God's word "the creaturely dignity of all things" and therefore the necessity of study for social justice.

Emmanuel Levinas says that for Judaism the goal of education consists in instituting a link between man and the saintliness of God, and in maintaining man in this relationship. It is a moral enterprise in which the fate of the world and our own human fates are inextricably linked. Not only is education ontological in that it requires us to be for the other, but it is moral in that its ends must be ethical—I think education provides us with a stance we might take in the world. We must find the lost objects and ease the burdens of others, and from captivity relieve all still enslaved.

Chapter 4

From Sinai to the Water Gate:
Curriculum Stories

I am finishing my sixth decade. Every third thought is of the grade.

Abraham Joshua Heschel in *Heavenly Torah* (2006) distinguishes between *halachah* and *aggadah*. The former is traditionally understood as the law, and the latter the stories and tales the Rabbis and others tell in the context of talking about the laws. They also represent two philosophical methods representing two ways of life and exegesis. "Halakhah is the yoke of the commandments; Aggadah is the yoke of the kingdom of heaven. Halakhah presents the letter of the law; Aggadah brings us the spirit of the law. Halakhah deals with matters that are quantifiable; Aggadah speaks of matters of conscience" (2). Aggadah holds that what can be measured does not contain the wonder of life; rather, "life's blessings are found 'in that which is hidden from view.'" Aggadah, the stories, are our answers to the question that the law too often raises but does not address. Aggadah, it is said, underlies even the law. If Halakhah represents the way in which we walk, then aggadah expresses how we feel. Heschel reminds us that what begins Torah is not the laws of *Leviticus* and *Deuteronomy*, but the *Genesis* stories of Creation, of the original first families, and the liberation stories of *Exodus*. Story, Heschel argues, lies at the foundation of Torah.

Stories, as do prayer and study, arouse in us wonder. The story, as is study, is a prayerful act. When we tell our stories, we acknowledge in public our sense of wonder and awe. Wonder, as Abraham Joshua Heschel teaches, is a radical amazement; wonder is a state of maladjustment to words and notions, the recognition of their fluidity. Wonder arises in the awareness of the world's glory that always exceeds our comprehension and our grasp. To our sense of wonder we respond with awe, and when we stand in awe, we acknowledge that even in the smallest particle there is meaning we can never fully understand. We tell stories about what we do not yet know. Stories derive from wonder. In our storytelling, we acknowledge how little we know, and in those stories we stand in awe at the complexities of our lives that only we, in part, can realize. There is always another story to tell; we tell more stories; we continue to learn.

A story is a moment of insight, and an opportunity for direction. Stories are what we have to narrate the sense of our lives. Our stories are our legacies. I would like to tell three stories about stories told. Curriculum is the story we tell our children. I would like to talk about some stories whose meaning, I think, is central to our understanding of education and curriculum. These stories, narrating events that are actually separated by almost seven hundred years, talk to us about the foundations of curriculum. These stories recount instances of public education and the imposition of standards; the first of these stories narrates the communities' failure to exhibit concern for their children even as they planned the children's future. The second story portrays tragically the results of the adult's moral and ethical recidivism with regard to purpose and learning, and the effect of that impoverishment on the children. Finally, in the third story, we hear about a remarkable tale of return—of teshuvah—of an educational turning toward the world to work for the world's redemption. The third story is a triumphant chronicle relating the people's acceptance of learning and study as the basis of ethical existence. This story, according to me, is the quintessential story of curriculum and of education.

The Children at Sinai

In neither the account of the Revelation in *Exodus* 19 nor in *Deuteronomy* 5 is mention specifically made of the presence or activity of the children. A remarkable educational event occurs that will dramatically alter their lives and yet, the children are, as it were, written out of the account. Were they not to be part of the covenanted community? Did not they, too, have to accede and receive instruction? I am concerned with the children: our public education has since its inception been concerned with the instruction of children. I am a teacher, and I have lived my entire life with children. There are, as Piaget reminds us, children all about us. The Sinaitic Revelation was a moment of absolute educational significance: it promised to impose standards and create culture. Revelation would teach the community beliefs and practices to be held in common, and the people—newly freed slaves—would become an ethical society. But during the accounting of this remarkable event at Sinai in which God reveals His self to the people, and offers the community as a whole a glorious future in exchange for a firm moral commitment—learning of and dedication to rigorous standards—the presence of the children is ignored. They were, it would seem, covenanted in their silence. I am certain that they were present—where in the desert wilderness could they have been sent so as to absent themselves from this felicity? They left Egypt with their parents, and the Rabbis tell us that it was the children who were sent to collect the Egyptians' gold and silver with which Israel left slavery. But in the narrative of the Revelation, the children are rendered invisible; they were spoken for, but they did not speak.

It is to my mind an egregious omission in the story. If Revelation is evidence of God's turning toward the human, the silence of the children in the narrative omits God's turn toward these most vulnerable children. If, as Heschel tells us, the indigenous quality of Revelation is to be found "in the creative fact of how the divine was carried into the concrete experience of man," the silence and invisibility of the children suggests the absence of the divine from their lives. What was the children's experience at Sinai? What did they see, and what could they know? At Sinai, the entire course of the children's lives was to be set by the events that would occur there; and yet no reference in the text is made to them. A whole lifetime of learning was set and lives foretold, and yet the children remain unseen and unspeaking. Perhaps it is that this is the lot of children—to have their lives set by the adults in the children's silence and absence. For the sake of my own children, I hope that this is not so; I would have them attain greater voice. Nevertheless, to describe this incredible occurrence at Sinai and to ignore the presence of the children startles. It would be like talking about the structure and purpose of a national educational system without the mention of the children for whom it would be designed and who would be in attendance!

I think that curriculum—what occurs in schools—is too often created without reference to the children; we wonder about the world into which they may go, and we wonder about the world from which they derive, but we remain unconcerned with the world in which they *live*. Perhaps we might look to the narrative concerning the Revelation at Sinai to consider what consequences such silence might have on the education of our children. Perhaps we might discern some meaning of curriculum stories from the events at Sinai. Here I would speak a few questions into the silence. This speaking is the Jewish practice of Midrash—the story that fills the space of the question.

Surviving Sinai

Where were the children? Of the presence of the adults at Sinai we are assured: the Rabbis say that when God commanded Moses to address the community to convey God's covenantal offer, God included in his address both the adult men and women present: In *Exodus* (19:3–6), God commands Moses:

> ³So shall you *say* to the House of Jacob [by which God means the adult females] and *relate* to the Children of Israel [by which God refers to the adult males]. ⁴You have seen what I did to Egypt, and that I have borne you on the wings of eagles and brought you to Me. ⁵And now, if you hearken well to Me and observe My covenant, you shall be to Me the most beloved treasure of all peoples, for Mine is the entire world. ⁶You shall be to Me a kingdom of ministers and a holy nation.   (Italics added)

The Rabbis even make meaning of the different verbs in the text to distinguish the manner of address to both the men and women: the verb *tomair*—to say—implies a mild form of address whereas the verb *v'tageed*—to relate—suggests a resolute form of address: to the men, God issues commands, but to the women God offers, as it were, proposals. But where in this great address were the children? It was a remarkable event, I think, and certainly one decisive to the children's future. God's revelation of God's self to the community at Sinai and the giving of the law was the event to which the delivery from Egypt was meant to lead: deliverance in this sense might be understood as not *out of*, but *into*. There were standards to be elucidated to and embraced by the community. The Covenant was intended to forge a community of high ideals and common purpose. The very quality of the children's lives was to be forever altered in this promise. But in the Biblical narrative, the children are not only invisible but silent as well. The people's commitment to the eternal covenant was made not only in the children's absence, but in their muteness. I wonder, into what were the children led? By what authority did the adults commit the children to a future in which they had no voice? Of course, it is just like us adults to obligate our children to *our* futures!! And at Sinai, as I will suggest, that future was sorely conflicted.

The children must have been there at Sinai, and they must have been children. Many of the children with whom I am familiar would have been alarmed at the frightening events on that awesome day. At Sinai, the sky darkened; thunder rolled, and lightning cracked. At such terrifying events as must have been the Revelation, my children often seek shelter from the storm. At Sinai (*Exodus,* 19:18) when "the mountain smoked all over,

since God has come down upon it in fire; [and] its smoke went up like
the smoke of a furnace, and all of the mountain trembled exceedingly,"
when the world itself seemed about to crumble, how much more so would
the children have experienced great alarm and sought some refuge. Even
were they actually present at the foot of the mountain, I suspect the children would have hidden themselves under the cloaks of their parents. The
shofar sounded without pause, not unlike the warning sirens at out home
in the Mid-West that urge retreat into the basement during dangerous
storms. From what I know of children, I think that at Sinai they would
have eagerly sought refuge in the basements, though in the desert around
Sinai there were, alas, none present. Perhaps God especially insulated and
shielded the children in that time of great awe and terror. I consider the
possibility that we are not so careful in the schools, though we do provide
for tornado drills and train the children in the event of fire. Today we
develop crisis plans for school shootings. But nevertheless, if at Sinai God
protected the children, then to what degree could they have been expected
to have experienced the Revelation?

You see, I cannot cease wondering about the children: I mean, I live
with two of them, I have spent my days for the past thirty-five years with
them; I was one myself. Where were they at the Revelation? Where was I?
The adults, we are told, "saw and trembled and stood from afar." The adults
would not come too closely; if they heard they did so from the distance. If
the adults trembled, how much more so must the children have shuddered?
What did the children think in the midst of this awesome sight? What did
they hear as the words were about to be spoken? What might they have
understood if, in fact, they heard anything at all? What could they comprehend about "a nation of ministers and a holy people?" I cannot imagine
that the children were not terrified. Indeed, might not the children have
experienced even greater fear than did the adults—after all, what could
they grasp about what was taking place? Even less could they comprehend
to what they had been committed: though they were, after all, part of the
declaration: "Everything that God has spoken we shall do!" If the adults
could hardly comprehend to what they had committed themselves, how
much more so must the children have been confounded by the events taking place and the vows made. I do not think the children could apprehend
the responsibility to which they had been obliged. Could the children ever
have realized the enormity of what was then occurring? Could they ever
comprehend that what they did not then know would forever change their
lives? Revelation, as does curriculum, has a life-altering effect.

And what curriculum at Sinai could the children have been responsible
for when, in fact, the children were not present or presented? How could
the children be held accountable for the promises of the adults? Of course,

it is often the case for children that adults commit them to a future to which the children have not themselves acceded, and in which the adults themselves show little faith or interest. It is also true that the children are often covenanted in their absences. After all, what else is curriculum in contemporary America but something to which the children are committed by adults in the children's silence and oft-times absence. Today, we speak regularly of standards externally determined that must be individually met by all students. We adults know to what they, the children, must be bound and committed. We adults confidently and imperiously determine what every fourth grader should know! We adults know what will make them great. Yet, at Sinai the children were bound to a covenant the adults themselves would not and did not keep. More's the pity for curriculum and for the children, I say. Though we expect that our children must know all of the state capitals, we adults take easy recourse to a map. Or we ask our friend. Often what we commit our children to learn, we cannot do ourselves. My father would say as he drew heavily on his unfiltered Chesterfield, "Do what I say and not what I do, don't smoke." If Revelation is the turning of God toward man, then soon, at Sinai, the adults turned away from God. Perhaps that adult turning from God derived from the same adult turning from the children. The absence of the children whose future was so bound to the events at Sinai offers insight into the meaning of this story—because what the children may have, but did not learn may be read into the narrative from their absence in it.

At Sinai, God commanded Moshe to teach the people to prepare to receive the Law, but there is no mention made of the preparation of the children to receive the Words. What were they prepared to receive? What could they have received without preparation? What did the people do with the children as the people themselves prepared for the events of that momentous third day? We know that God commanded the people to wash themselves and their clothes, and to remain celibate—to purify themselves as it were. Did the people scrub their children and demand that they sit on the doorstoop in their newly laundered clothes; did the adults order the children not to move or get soiled? For what did the adults tell the children they were awaiting? For how long did they tell the children they must wait? Perhaps from the point of view of the children, in the account of the Revelation we have one inspiration for Beckett's *Waiting for Godot*. There is, in fact, no mention of their preparation. And when we prepare children for school, I wonder, do we tell them for what they wait? I wonder, sometimes, if it is not with our own childhoods that we prepare our children.

If the children, as I suspect, weren't present at the Revelation as even the adults were present, then by what authority should the children be held responsible for the adults' actions—in either the acceptance or the disavowal of the covenant? How can we plan our children's futures when we absolve responsibility for our own, as the adults would soon do at Sinai? If the children perceived no revelation and received no instruction of it, for what can they be held responsible? And if the children were not there, what could they have learned? I think that the narrative of Revelation clearly suggests that children were taught neither the meaning nor particulars of the covenant. Rather, the narrative speaks of an incredible failure on the part of the adults to either practice or to teach. In the almost immediate failure of the adults to maintain faith at Sinai, in the making of the Golden Calf, the adults sentenced their children to forty years of wanderings in the desert; the adults, themselves unable to sustain their commitment were wholly unable to help the children learn. The Revelation was to commit the children even in their silence to a way of being and an order of living. I daily wonder by what authority we adults obligate our children to our adult ways of life? This is, after all, the traditional business of schooling and curriculum, isn't it—the deliverance of the requisite culture to the next generation? We debate these curricular issues daily; they lay behind the issue of national standards and multicultural education. Again, we might perhaps look to the experience of the children at Sinai to gain some insight into this debate, for there the adults acquired substance and could not teach it.

If Sinai is the beginnings of formal curriculum, as I have suggested, then it is a troubled beginning for schooling and curriculum, indeed. Our children's lives depend on greater concern and rigor. Perhaps, we might look to the silence and absence of the children to understand the significance of the events surrounding Revelation to education and curriculum. As we adults plan the children's education—as we organize curriculum—we would do well to keep in mind where they are—and hence, what they might know and what they might be yet earning—for even in their absence their presence must be recognized. Indeed, they learn anyway. Perhaps this is one meaning of the silence in the biblical narrative regarding the children. Silencing the children denies the future and the people's future was situated in the empty wilderness.

I would suggest that though the Revelation at Sinai might be understood as the first official reference made to issues regarding curriculum—it was after all, a moment of the elucidation of formally defined standards that were to be communally learned and kept—the children were not only not present at the Revelation, neither did they receive the learning meant for them. The Revelation was meant to be a learning experience: "if you will hearken, yea, hearken to my voice and keep my covenant...you shall be to me a special people." There would, it would seem, be a test. But the adult community failed this test miserably—they did not hearken and they did not keep the covenant—and this failure condemned the children to years of wandering in the desert. In all of this pedagogical talk, there is no mention made of the presence or experience of the children. I wonder if, perhaps, the adults were not too caught up in their own experience at the Revelation to be concerned with the children. But as a father and an educator I wonder in what and by whom were the children instructed? Their apparent absence does not mean that they were not affected nor that they did not learn.

Children at the Revelation

I would like to propose that curriculum derives its authority not from what is heard, but from what is felt. Curriculum draws its power from a belief in the integrity of the story. The story is, of course, curriculum—what we tell our children—and I think that story, whatever it may be, must promise a viable future to inspire faith in it. But at Sinai, I cannot imagine that what the children felt was anything but fear and trembling. The story of the Revelation from the point of view of the children narrates terror, weakness, and finally, failure. The story narrated speaks in absences—absence of faith, of vision, and of hope. Perhaps that is why the children are absent from the narrative; they represent the future that the adults could not envision. The story at Sinai speaks of the past and denies the future; it is a story, finally, not at all about the children. It is not of their education—their lives—with which the story deals. Rather, the events narrate a return to idol worship and a denial of the unseen God. It is rather, a story of the *failure* of education. Though the Revelation was meant to teach, the people lacked the capacity to learn. *They did not have faith enough. They felt fear.* From that fear they implored of Aaron to "Make us a God who will go before us." I suggest that it was at this idolatrous act that the children were present; it was this momentous event that the children saw; it was from this occurrence that they finally had instruction and from which they learned.

The Golden Calf

The building of the Golden Calf gives evidence that the Israelite nation that camped at Sinai were tried and did not succeed at the trial. The children, I am certain, witnessed this failure. From this failure derives, perhaps, another curriculum lesson. If the absence of the children from the narrative suggests that they were not even present at the Revelation, or at least were not conscious *of* the event *as* an event, if they could not possibly have apprehended even the smallest part of what the adults apprehended, then what they might have understood from the first experience—the incident of the Golden Calf—at which they were certainly in attendance, is of primary concern to me here. It was the first moment after the Teaching. Inevitably, the children learned from the adults' failure; inevitably, they were schooled here. In education, we refer to such learning as the product of the hidden curriculum. Indeed, because these learnings derive from silence these forms of curricula are powerful: left unspoken, they cannot be refuted or even discussed.

God had known that the adults at Sinai would fall into idolatry. Maimonides (1963, II.498) says:

> Know that the aim and meaning of all the *trials* mentioned in the *Torah* is to let people know what they ought to do or what they must believe...It is in this way that the meaning of *trials* should be understood. And it should not be believed that God, may God be exalted, wants to test and try out a thing in order to know that which God did not know before. (Italics in original)

God knew that the people would fail at Sinai and would raise up other gods. The events at Sinai, unlike our instruments in schools, were not meant as a test to measure learning. Rather, says Maimonides, the purpose of a trial is to teach. As should be the case in any great classroom, tests ought to be pedagogical instruments. Maimonides (498) says, "Accordingly, the notion of a trial consists as it were in a certain act being done, the purpose being not the accomplishment of that particular act but the latter's being a model to be imitated and followed." That is, the accomplishment of the test serves as the model for behavior—as in, say, the ten trials of Abraham or Jacob's wrestling with God. The test is not proof of learning; rather, the tests serve the purpose of the learning itself. In the desert at Sinai, in the hours and days subsequent to the Revelation, the people failed this test miserably!! They did not believe; they had no faith. Now, as a teacher I know of what Maimonides speaks: there is learning from failure as well—though such lessons are often more difficult and painful and often have greater consequence. The building of the Golden Calf as a public display of lack of faith condemned the nation to forty years of wandering in the wilderness.

I would propose that for the children the meaning of Revelation occurred in the incident of the Golden Calf, the first event at which they must surely have been present. In the incident of the Golden Calf, the children might have come to understand the absolute necessity for boundaries and for borders that faith in a future provides: the unitary God asserts singular, discrete, and uncompromising commandments for present behavior. The people readily and unthinkingly acceded. Maimonides argues that at Sinai the people heard only the words "I" and "thou shalt not"; after that they heard only God's voice but not the distinct words. And if the adults heard so little from the single unitary God, I do not think that the children apprehended even the words "I" and "thou shalt not." I think the children trembled in the metaphorical basement. Or they hid themselves in the folds of their parents' garments with their hands held firmly over their ears. However, at the incident of the Golden Calf, the children existed: I think they helped gather the jewelry that would be crafted into the molten idol. Who else would have been sent back to the tents to gather the riches with which the Israelites left Egypt? "I" asserts the presence of God; faith provides the frame in which behavior commanded by that unseen God might occur. But "thou shalt not [have other gods beside me]" must first be transgressed so that the idea of the singular and omniscient God may be ultimately understood and accepted. Faith here gives structure to existence. Faith argues the morality of learning. The unseeable, unitary God issues a single structure of values based in a consistent value system: *if* you accept my commandments, you shall be to me a nation of priests and a holy nation. The value of the commandments and the values that inhere in them depends on faith in the God that enunciates them. God's presence must be first acknowledged before God's commandments can be learned. Upon such faith the laws that will be given voice may be understood and accepted.

But the idolization of the Golden Calf revealed that in this trial of faith the adults of the community miserably failed. They did not have faith, and the children served as witnesses to that failure. I think that what the children saw and heard was, indeed, revealing. Of the events of the Revelation at which the children were not present, the children would ask of the adults, "What happened? What did these events mean to you?" And the adults would respond, "We have faith in God. We are to become a nation of ministers and a holy nation." "What does that mean?" the children might ask. And the adults might respond, "We do not yet know, but we yet have faith. We will wait." But as the people danced about the Golden Calf the children would ask, "What do these events mean to you," and the adults should have answered, "We lost faith. We did not trust. We did not believe. We could not wait. We had no resources to control our fears. We could not contain our desires." "Ah!!" the children might respond.

When Moshe descends the mountain fully expecting to deliver God's word to the awaiting community, he hears and observes the frenzy of the idolatry, the consequences of the people's lack of faith. In his fury, Moshe throws down the tablets and castigates the people. Perhaps it is at this moment of the people's painful contrition, and in the face of Moshe's anger, that the real meaning of the Revelation for the children might be found. For suddenly, the events of the exodus and the desert wanderings and the occurrences at Sinai become narrative. It is a human story the children perceive. It is a story of the trial and the failure of the people to withstand doubt and uncertainty. It is a story of the failure of faith. Until that moment the people have been led; they have been actors in and not authors of the narrative. Here, awaiting Moshe, the people are asked for the first time to do something for themselves—to wait the pronouncements of the unitary God—but in order to wait they must first have faith in God's existence. The Golden Calf is evidence of a lack of belief, an absence of faith. The people would not learn. The children were present at this failure.

The story we tell our children concerns the world which may be known by them. Curriculum is the story we tell our children. When the adults promised to do what God said before God had said anything, the child might have responded that the emperor was not wearing clothes. The children might have asked what exactly they were expected to do. Children are immensely practical. Perhaps that, too, is why the children are excluded from the narrative of Revelation. They would have raised precisely the issues the incident of the Golden Calf would illuminate—the issue of faith. But the adults' behavior demonstrated for the child not merely that the emperor was wearing no clothes, but that there was, indeed, for the adults, no emperor! Ah, but Moshe's anger—and I suspect, Aaron's reluctance to participate in the idolatry, confirmed for the children God's presence. And then there were the stone tablets that Moshe carried, albeit briefly, on which the actual words were written. The children would have wondered where Moshe had gotten such an incredible object. And so the revelation in the incident of the Golden Calf taught the children that action derives from faith; faith is the frame. Without a frame there is no story, but only the frenzy of the moment. The adults had no faith. If the Israelites were to be a holy people and a nation of ministers—if the community was to possess a moral basis upon which learning will derive—it must first begin with a faith that there is something beyond us to which we aspire. There is meaning beyond what we can know. That is the definition of awe, but it is as well the idea of wisdom. At Sinai, the people's abandonment of faith revealed the value of faith. The people at Sinai lacked a frame—they did not themselves know about faith what the children would learn—and without boundaries the adults grew frightened. At the dramatic events at Sinai the children must have been frightened, and in their fears looked to the adults. But the adults could offer the children little that would sustain. The adults had nothing to offer. I wonder upon what faith we ground today's curriculum of national standards.

Toward the Water Gate

What curriculum existed in the forty years of desert wanderings? There is no sense in Torah that God's word was studied during these years; I read not of the establishment of *cheders* and *yeshivot*. Only after the conquering of Ai in the twelfth century BCE (*Joshua,* 8:34–35) does the text finally refer to the reading of the book of Moses. "Joshua read to the entire congregation of Israel, the women and the children and the converts that walked among them" the words that Moshe had commanded. Indeed, until that moment, quite the contrary is recorded in Torah—we read of the grumbling of the people expressing their distrust of God, and we read of Moshe's repeated intercession to prevent God's annihilation of these stiff-necked people. We read of the people's repeated rebellions. Again we recall that Maimonides argues that a trial is a test—we must again acknowledge that the people did not pass this test. If the desert made them strong, it did not make them wise. After Moshe returned with the second set of tablets and told the people all that God had commanded him to speak, what in fact did the people actually do? They grumbled. If at Sinai the value of faith was revealed, then the experiences of the desert gave little evidence of faith.

I wonder how the children were instructed prior to Joshua. After all, we know the history of the Israelite nation is a troubled one of idolatry and rebellion. What story were the children told during the desert wanderings? What happened to the Revelation during these years? Curriculum is, after all, the story we tell our children. We must tell the children something: what were they told? Children demand to know. Four times children are referred to in the Torah asking questions: they are the origin of the four children referred to in the Passover Haggadah. They look about them at the behaviors of the adults, and the children simply want to know what is going on. They don't yet understand!! It is to behavior that the children refer: why are you behaving this way, the children ask. What does all of this mean? What is happening? Are we going to die? The story of the four children in the Passover Haggadah is a reminder that children yet require explanation; they do not ever accede without cause and consistency. Thus, even if the children were not only in attendance at Sinai, but attentive as well (which as I have said, I am doubtful), the children still require rationale. They ask: "What are the statutes, the laws and the ordinances which Adonai our God has commanded us? What is the meaning of this ritual to you, so that I may in my understanding accept it?" If God tells the community that they are to be a holy people and a kingdom of priests—then my child first wants to know what is holy and what is a priest. And

I must tell a story to explain it. It is the obligation of the adult to narrate. Interestingly, four times the parent is instructed to answer the child's questions regarding the customs and practices (*Exodus,* 12:26–27; 13:8; 13–14; *Deuteronomy,* 6:20–21), but in fact, the tale told is of the exodus from Egypt and the sacrifice of the paschal offering, a practice that was abandoned during the desert wanderings. For the most part, there would have been no practice for the child to question, and as I have suggested, no faith to support practice. It is our faith and our values on which the story will be based—though at Sinai, as I have said, there was only the verbal commitment and none of the faith—indeed, at Sinai, the faith did not abide and could not sustain. Without vision the people perish, the children learned long before Isaiah spoke.

The story we tell is the way of life we wish to give the children; the story contains our hope of a future. At Sinai, the future did not occur to adults though the presence of the children bespoke it. And the absence of the children in the account of Revelation bears witness to the adults' lack of vision and of faith. Without faith the people perish. The narrative of the Revelation at Sinai and the consequent wanderings in the desert reveals to us the necessity of faith; at Sinai the adults cried, "Everything that God has spoken we shall do." But soon they asked, "What will God say? And when? I need something to do now," they demanded!! The ability to wait is faith. I speak not of a quietism; I speak of a wise patience. I speak of awe. If the children learned the value of faith at Sinai, they were given nothing there to hold onto. The longer they wandered in the desert the more distant from the story they moved.

You see, finally the world needs explanation. Even for those who do not ask we are required to provide answer regarding our behaviors. It is, of course, possible that the child who does not know how to ask is actually awestruck, but the answer offered still must concern the story. In the wanderings in the desert what did the stiff-necked grumbling people tell their children? What did the children grown to be adults in the desert know to tell? If curriculum is a story we tell our children, then it is a sordid story told to the children in the Sinai desert; it is not that of how to be a holy people. I think the adults were too occupied complaining about their existence to be concerned with the Sinaitic story.

It is a hint to us that as we adults bemoan our conditions we neglect the business of our lives. We neglect our children. It is a cliché to say that a society must educate its children, but to do so properly it must first be prepared to answers its children's questions. We must respond to their whys and their whats—and the story of the Revelation at Sinai suggests that our answers ought to be based on faith. "You shall be to me a nation of priests and a holy nation." But in the desert and amidst the hardships and trials, I suspect such faith was in short supply and that the children's questions were ignored and/or dismissed. They were apprehended, as children's questions too often are, as a nuisance. Or the answers proffered had to do with expediency and not faith. If there was curriculum in the desert—and, of course, there is always curriculum in deserts and in schools—then that curriculum was informed by no belief. It may have been practical—but it was not enduring. I think the children were the guardians of faith, but they had no means to sustain it nor to act on it. The adults were too preoccupied. Perhaps that is why there is no mention of children at the Revelation.

Curriculum at the Water Gate

What happened to Revelation? I mean, we possess a text and it is studied, but during the forty years of desert wanderings that text was ignored, abandoned, and discarded. What happened to the text and the people's relationship to it, their knowledge of it, during the long years of exile? The commentary to *Deuteronomy* states that "Had Shaphan not arisen in his time, and Ezra in his time, and Rabbi Akiva in his time, would not the Torah have been forgotten in Israel?" In tractate *Succot* (20a) of the Babylonian Talmud and in *Sifre Deuteronomy*, the Rabbis say that "When the Torah was forgotten in Israel, Ezra came up from Babylon and established it." But, I wonder, if the Torah had been forgotten, then what text did Ezra establish? What text did Ezra remember? In what language did he remember the Torah? In what language did he speak? We read that through Ezra even the writing of the text was changed, as it is written: "And the writing of the letter was written in the Aramaic character and interpreted in to the Aramaic [tongue]." What Torah contains may be considered a sacred trust, but during the wanderings in the deserts and in the idolatrous years that followed, and certainly in our own deeply troubled times, Torah's sources had become distant, unstudied, forgotten, and therefore, problematic. What happened to the text over the years in which it was abandoned and not cultivated, in which it was ignored and not kept, and in which it was cast aside and not studied? How can I understand the effects on the text of the many years of idolatry and moral decline in the community? If the Revelation imposed standards, then what does the narrative surrounding the Golden Calf and the years of desert wanderings and grumblings and idolatries suggest about the integrity of those standards? If curriculum, as I have said, is the story we tell our children, then the biblical narrative recounts the story of the Golden Calf, of the people's grumbling, and their abandonment of the commitment. The people, the story says, had no faith. "They have strayed quickly from the way that I have commanded them," God tells Moshe even while Moshe is still atop Sinai. The people, as I have said, did not withstand the trial. They had to learn, even if they did not do so. Nor did their contrition ever prove lasting: the adults continued to murmur and to grumble and to rebel. If the people did not practice the words, and if they certainly did not teach them to the children, then I do not think they could have been subsequently remembered nor taught nor learned. Certainly, the preservation of the words is doubtful. When Ezra reads the people the Torah of Moshe at the Water Gate he reads a text, but I am curious what text he read.

Upon the return from the Babylonian exile, the people ask to be given the scrolls of Torah. The event mirrors, as it were, the Revelation at Sinai with at least two significant variations that I want to address here. First, of course, at the Water Gate the people request the scrolls rather than having them offered as part of a vague commitment, and second, the entire scrolls are read to them at once. And the scrolls are discussed. We read:

> Then all the people gathered together as one man [sic] at the plaza before the Water Gate and they asked Ezra the scholar to bring the scroll of the Torah of Moses, which God had commanded to Israel. So Ezra the Kohen brought the Torah before the congregation—men and women, and all those who could listen with understanding—on the first day of the seventh month. He read from it before the plaza that is before the Water Gate, from the [first] light until midday, in front of the [common] men and women and those who understood; and the ears of all the people were attentive to the Torah scroll. Ezra the scholar stood on a wooden tower that they had made for the purpose; next to him stood Mattihiah, as well as Shema, Anaiah, Uriah, Hilkiah and Maaseiah on his right; and on his left, Pedaiah, Meshael, Malchijah, Hashum, Hashbaddanah, Zechariah and Meshullam. Ezra opened the scroll before the eyes of all the people; and when he opened it, all the people stood silent. Ezra blessed God, the great God, and all the people answer, "Amen, Amen," with their hands upraised; then they bowed and prostrated themselves before God, faces to the ground...They read in the scroll, in God's Torah, clearly, with the application wisdom, and they helped [the people] understand the reading.   (*Nehemiah*, 8:1–8)

This is the first public reading of the Torah scrolls in at least seven hundred years; indeed, it is almost the first time the scroll has ever been read by request. I might suggest that it is here at the Water Gate that the people hear the whole text for the first time. As education requires engagement with the text, then this might also be the first instance of the community engaging in public education. Second, the meaning of the text is mediated at the Water Gate by human explication—as the Talmud will say, the statement "it is not in Heaven" suggests that meaning resides in human endeavor and not in heavenly decree. Indeed, perhaps that was exactly the problem at Sinai—too much was in Heaven and there was too little for the humans to do. At Sinai one man, Moshe, bore the entire burden of the reception of the text and then the communication of the words; the people had little to do and less to consider, and they easily slipped into idolatry. Nothing existed to sustain their faith—they possessed nothing but vague intention. Perhaps all futures require texts that must be read and explicated. But here, at the Water Gate, the responsibility is shared by a rather large community each of whom are specifically named—they are individuals and I suspect, possessed of differing intellectual proclivities. They do not think alike. There will be, I anticipate, a variety of interpretations and opinions. The people will hear a multiplicity of texts.

What is the status of that text read at the Water Gate? What relationship is there between this text and that received at Sinai? In the Avot of Rabbi Natan there is this story:

> Deuteronomy 29:28 states that "The secret things belong unto the Lord our God; but the things that are revealed belong to us and to our children forever." But over the letters that spell out the words "to us and to our children" and over the first letter in the word "forever" there is a dot. Why? Because Ezra said, "If the prophet Elijah comes and asks me, 'Why did you write it [the Torah] thus?' I will reply, 'But I did put dots over the letters [to indicate my uncertainty about the text].'"

The text is imperfect. In this midrash Ezra concedes his own uncertainty regarding the soundness of the text; nevertheless, he is prepared to study it, to restore its integrity and to live happily with the ambiguity. His work does not deny the ambiguity but depends on it. It is probable, says David Weiss Halivni, that the Torah that Ezra read at the Water Gate seven hundred years after the Revelation was "an uneven and composite text—sacred and inviolable as the tangible sign of Revelation, but in need of emendations and additions when brought to the realm of practice." The text read by Ezra was, nonetheless, the same Torah that Moshe had received at Sinai: "It has been taught: R. Jose said: 'Had Moses not preceded him, Ezra would have been worthy of receiving the Torah for Israel. Of Moses it is written *And Moses went up into God,* and of Ezra it is written, *He, Ezra, went up from Babylon.* As the going up of the former refers to the [receiving of the] Law, so does the going up of the latter…'" But for each the Torah is clearly a different text. Ezra's Torah is equivalent to Moshe's, but is not the same. Both texts are the product of God's turning toward the human being, but the text Ezra and his colleagues read requires repair and resolution.

Unlike at Sinai where the people stood and listened from afar, at the Water Gate, the people come forward to hear the word that they had themselves requested be read to them. They are prepared to engage with the text. They perceive their own needs. At the Water Gate, Ezra and the leaders who stood with him "read in the scroll, in God's Torah, clearly, with the application of wisdom, and *they helped [the people] understand the reading*" (italics added). The people desired to learn, and Ezra and those who stood with him became the people's teachers. That is, Ezra and the leaders who stood with him invented exegesis—textual study—they created as it were, curriculum. Though the Torah was not originally given through Ezra, the Torah that Ezra reads at the Water Gate is equal to the Torah that Moshe received at Sinai. Exegesis becomes the means of the text's voice. The Torah Ezra reads helps the people understand and invents hermeneutics. In his reading Ezra restores Revelation.

If the story of Sinai speaks of the necessity of faith in faith's absence, then the story of Ezra speaks of faith's renewal and restoration. The story of Ezra narrates a construction of contemporary curriculum. Ezra read to the willing people the story. Ezra read to the willing people *their* story. Ezra helped the willing people *understand* their story. Curriculum is the story we tell our children. The story is all we know. It begins in faith—that meaning exists and it is consistent and entails responsibility to what is always beyond our grasp. We must first have faith. And as education begins in faith, it is realized in an earnest desire to learn: at the Water Gate the ears of all the people were attentive to the reading of the Torah scroll. They wanted to understand and their teachers helped them to understand. This Revelation will be life—changing forever as well. Curriculum must enact faith; it must augur the future. Faith is a beginning, but faith must be continuously and willingly restored and renewed. The Sinaitic experience is a singular story to tell our children, but the adults lacked the faith that they could endure Revelation. Or perhaps we might say that they lacked the faith to make Revelation endure. How much more so must the children who had been absent from Revelation have been capable of that achievement. Whatever education they received in the desert—and I am doubtful of its substance and integrity—it was not derived from Revelation. Indeed, as I have said, we read that much of their curriculum consisted of grumbling and rebellion. During the desert wanderings I do not think much of the text was ever taught to the children. Faith could neither be sustained nor enacted. Revelation was abandoned and its meaning lost.

By definition, Revelation occurs only once. But the substance and significance of Revelation must be continually rekindled. Faith must be renewed and sustained, but how and by what might it be so? I think faith must be enacted—practiced—or it is empty and meaningless. First, however, our story must be communicated—it must be taught. Over time the story varies with the people who tell it and to whom it is told and the conditions under which it is narrated. We imagine a means to ensure the truth of the narrative, but the story of the Torah text suggests that no text remains immaculate. Thus, we must discover an expertise to create it. There must be a means to restore and then renew Revelation. Faith is perhaps enacted and given substance in both this restoration and renewal. Curriculum must give substance to a faith upon which it is founded.

At the Water Gate, the people requested that the story be again narrated. Curriculum, as I have said, is the telling and retelling of the tale. What was revealed was done so once. But the events and consequences of the Revelation might be forever recounted and understood. And therefore, renewed. That process is the substance of study.

I am, as I have said, an educator. I tell stories. I am also a father. I tell stories. I am concerned with the integrity of the story told. This is an issue of standards: what is important to learn. I am concerned with the integrity of the text, or rather, I am concerned with the integration of the text. Because we can never know fully—until Elijah comes, perhaps, the complete honesty of the text, I wonder how will any child comprehend this text? What will they understand of it and from it? How will they be able to read it? What reasons should they have to read it? The children must be instructed that this continuous activity of renewal, of textual study, is central to their lives. Their faith must be maintained and renewed in order that they might continue to have faith that there is something beyond their present time to which they should strive. They will continue to study the text. I wonder, will their Ezra tell them *what* to understand, or will their Ezra *help* them to understand.

Ezra's text was not the exact text as that revealed at Sinai though perhaps, it was the same text. Education begins in faith and is sustained in textual exegesis. It is the business of exegesis—which Ezra institutes at the inception of Torah reading—to recover the words, to restore the meaning to the words. As Halivni (1997, 83) says, "Once the nation had embraced a book, no need remained for the admonitions and the visions of the prophets. Interpretation took the place of Revelation." Study will make of the people a holy nation. Restoration of Revelation becomes an educational endeavor. Ezra ensures the continuation of Revelation by the institution of human study. Ezra, I think, introduces our modern notion of curriculum: textual exegesis. Curriculum is created. Curriculum creates. It is with curriculum that children spend the formative first twenty years of their lives. Curriculum is how we adults aspire. It should also inspire.

I do not think the children were actually present at Revelation, though perhaps they learned from it the value of faith. But at Sinai the children probably derived little substance. However, the renewal of Revelation at the Water Gate revealed to the people and to the children the purpose, the value, and manner of study: engagement with the text to construct its meaning in order to become a nation of priests and a holy people. Study must always be founded on faith; study embodies our faith in the future. Study is the engagement and renewal of texts. I think that both faith and exegesis are issues of curriculum. God knew the people would build a golden idol. Why did God permit the idolatry of the Golden Calf? Because Revelation afforded humankind the opportunity to change itself—to learn—but that change had to come from humankind itself. The quality of our lives derives from our willingness to learn: from our education. The Sinaitic Revelation taught that God demands something of us. We must maintain our faith in what remains always beyond us. The events at the Water Gate teach us how to discover what God demands. It is to study. We must continue to renew our texts and restore Revelation. Education is our means out of the desert. While we study, there is no desert. And this study built upon faith is the substance of curriculum.

Part 2

Every Third Thought Is of the Brave

Chapter 5

It Must Suffice

I am finishing my sixth decade. Every third thought is of the brave.

To be a teacher, I think, is to be brave. Each day teachers make decisions small and large, and remain attentive despite the myriad distractions and mind-numbing environments in which they work. During the course of each day teachers assert their wills, which is to say, to require by the constant effort of their behaviors that students think, and to provide relevant material with which students might continue to engage in thought. Theirs is a formidable task. William James sought for our lives the moral *equivalent* of war, and argued that the martial type of character, as already evidenced in the education of doctors and priests, and characterized, at least, by "strenuous honor and disinterestedness," must continue to be bred in all if our society is to constructively advance. The country must educate our citizenry for peacetime service just as we educate our army for war. James writes,

> Such a conscription, with the state of public opinion that would have required it, and the many moral fruits it would bear, would preserve in the midst of a pacific civilization the manly virtues which the military party is so afraid of seeing disappear in peace. We should get toughness without callousness, authority with as little criminal cruelty as possible, and painful work done cheerily because the duty is temporary, and threatens not, as now, to degrade the whole remainder of one's life.

It must be the *teachers* who are responsible for that learning. It is their *teachers* who would educate James' doctors and priests. Teachers are brave not because they stand "on the front lines" nor function "in the trenches," but because they have accepted the discipline explicit for the moral equivalent of war. Teachers serve as moral exemplars as a result of the very nature of their activities. Simply put, teachers think, and promote that activity in students. In his *Talks to Teachers* James (1962, 91) writes, "If, then, you are asked, *'In what does a moral act consist* when reduced to its simplest and most elementary form?'* you can make only one reply. You can say that *it consists in the effort of attention by which we hold fast to an idea* which but for that effort of attention would be driven out of the mind by other psychological tendencies that are there. *To think* is the secret of will, just as it is the secret of memory" (italics in original). This thinking is not a simple activity, but one that requires sustained attention and effort, and which becomes, therefore, an act of bravery.

Teachers should not only be brave, but must help others achieve bravery. Despite the lack of public support or sometimes, even attention, despite the steady calumny to which teachers are subject, the teacher daily engages with the individual and collective products of social and political pasts, and attempts from them somehow to create a viable future. James writes, "Thus are your pupils to be saved: first, by the stock of ideas with which you furnish them; second, by the amount of voluntary attention that they can exert in holding to the right ones, however unpalatable; and third, by the several habits of acting definitely on these latter to which they have been successfully trained" (91–92). I have argued (Block, 1998) that education is redemptive, and have more recently (2004) suggested that study is equivalent to prayer. Indeed, I claimed that to stand in the classroom is to live in awe and wonder. In the last chapter, I have even suggested that study prepares this world for the messianic era. James termed this future as socialistic, but regardless of its exact nature, that future promises redemption, and it is the teacher who facilitates this movement through her daily effort. Bill Pinar (2004, 61) writes that "Despite our professional subjugation by politicians seduced by "business" thinking, with its obsession over the "bottom line" (test scores), we must not succumb in spirit. We must remember that education is not a business, that it cannot be measured by test scores, that it is too important to be left to either politicians or parent." This temerarious stance in the world is not the solitary and unusual act of heroism for which Congressional Medals of Honor are awarded, but the daily effort to keep on keeping on.

In our society, the teacher is too often portrayed as an incompetent fool, a brute bully, an effete social functionary, or an amusingly powerless idealist. Each of these assigned roles serves the reigning perspective on education: an arena in which the uneducated are taught by the incompetent or the ineffective. The public portrayal of teachers in the news media, on television, and in the films does not usually flatter the profession, nor does it often portray the reality of the situation. One of the more recent portrayals that addresses the teacher's work realistically is *Freedom Writers*, the movie based on the real experiences of Erin Gruewell, played in the film by Hilary Swank, a dedicated new teacher in a "voluntarily integrated school" that became (especially after the Rodney King fiasco) an "involuntarily segregated school." Gruewell is portrayed as sufficiently brave; indeed, she is portrayed as a dedicated and effective teacher in a situation that obviously demands bravery and discipline. However, the film also points out a serious disconnect between the public demand on the teacher and the actual work the teacher must do. Gruewell is portrayed as exemplary and effective—*as brave*—but her portrayal contradicts every demand made on the teacher today as a result of the heinous *No Child Left Behind* Act. In this way, we permit Gruewell to become our mad and idealist Quixote, stretching the image of the perfect teacher without having to think very hard what connects her work to the daily lives of all of the other teachers in the public schools today. She is portrayed as the exception to every rule; in our admiration of her courage, we are led to define the cravenness (or simple effeteness) of the others by her example. She is the hero raising the second flag on Iwo Jima, and it is *her* image in the paper that sets the standards for the profession, or characterizes the rest of us as insufficient.

Despite my pique, I very much enjoyed *Freedom Writers*. I think the film says about teaching what I have learned about the profession in my thirty-five years engaged in it, and specifically over the past several years as I have, in my writing, come more and more to articulate that position: to teach is to assume an ethical position in an immoral world. To teach is to be a prophet in a degraded world. To teach is to not suffer silently, but to suffer nonetheless. To teach is to change the world paper by paper. Erin Gruewell was a wonderful teacher—I am not certain what she does presently, though the film suggested that she works now at California State University—because she was concerned with the lives of the children now, rather than their future lives that might never occur. Without ignoring the world outside the classroom, she attempted to create a safe place within it, and to offer her adolescents robbed of their adolescence, some place of relative peace to make some sense of their lives outside this artificial and temporary haven. In this impossible situation, she did not change the world, but she did manage to affect not a few of her students. It is their writings (and lives) that inspired this film. I wonder then, if she ought to be less the focus—though she is the hero, she is not alone in her heroism. Her complexities are too known, and those of her students less revealed, perhaps. She is too often in the center of the screen image. And though her students are not flat stereotypes, they are less than full characters.

Yet, though the film suggested that Erin Gruewell was a wonderful teacher—depicted as a incredibly effective, life-changing teacher, in all the best senses of that term—it was quite clear that there was *nothing* we could teach in our schools of education that could prepare our students to practice as did Ms. Gruewell. That is, how she was trained to be a teacher was never raised in the film as an issue: it was her own personality and particular circumstances in the classroom that made her a great teacher, defying the system as she did. That is, were we to offer a pedagogy of rebellious behavior and brash action, we would all soon be fired. Actually, some of us in higher education (whatever that descriptive term has come to mean!) know that there is indeed, something we can do to prepare more Ms. Gruewells. But this curriculum will sit well with neither administrators nor politicians. This pedagogy may suffice, they would say, for "the unteachables," but it will not do for those who are certainly college bound.

However, it is clear that if we do not change our educational focus for preservice teachers, we will perpetuate the world the Ms. Gruewells of this world must defy. So, though the film will make a great deal of money, and columns will be written about the courage of such teachers, in fact, there is nothing in our society in place to support either the Ms. Gruewells of this world or the educational structures that could produce more like her.

And of course, for a $27,000.00 salary, no one should have to forgo her life for so much effort and so little respect. The absolute nerve of a society to pay so little to those who care so well for their children!

But it is a story, and we teachers must pay attention to the stories we tell and the stories that are told. Indeed, here is another story:

> Whenever the Jews were threatened with disaster, the Baal Shem Tov (his name means *Master of the Good Name*) would go to a certain place in the forest. He would there light a fire and say a certain prayer. Always, a miracle would occur and the disaster would be averted.

The Baal Shem Tov (1698–1760) is a legendary figure in Jewish history. He is often considered a founder of Chasidism. In his youth, Israel Eliezer, who became the Baal Shem Tov, studied and lived rather simply. But in 1734, when he was thirty-six years old, the Baal Shem Tov revealed himself as a scholar to the world, and began to practice as a *rebbe*. He settled in Talust and rapidly gained a reputation as a holy man. Rabbi Israel Eliezer became known as the Baal Shem Tov—Master of the Good Name—because of his ability to perform miracles and to heal the sick. Later, he moved to Medzeboz in Western Ukraine, where he lived for the rest of his life. Interestingly, the Baal Shem Tov wrote down none of his teachings. All of his wisdom is known through the writings of his disciples who wrote down the Baal Shem Tov's sayings and stories (http://www.jewishvirtuallibrary.org/jsource/biography/baal.html). It is interesting to note that Rabbi Nachman of Bratzlav, the Baal Shem Tov's great grandson, also left no writings, though his famous stories were transcribed by his disciple, Rabbi Nathan Sternherz.

I think that it is comforting to have someone so perfect, wise and righteous, and who is capable and willing to assume all of our burdens and responsibilities. Who can go to the certain place in the forest, light a fire, and say a certain prayer. Who can prevent disaster. I suppose that such a personality might not absolve us of our daily obligations, but could serve as a safety valve in times that are communally beyond our control. The Baal Shem Tov could do it all. Or at least, so we are told. Once, Israel hired himself out as a teacher's assistant, and would lead the children to and from school through the woods singing the songs he had taught them for their lessons. They were beset on by a monster, and the children became very frightened; their parents refused to allow Israel to walk with their children any more through the forest, but Rabbi Israel Eliezer, soon to become the Baal Shem Tov, went from house to house asking permission from the parents to allow him to continue to lead their children to school, and promised protection to shepherd the children through the forest. The next time he walked with them, indeed, the werewolf attacked again, but Rabbi Israel Eliezer struck him between the eyes and killed him. The next morning the sorcerer was found dead in his bed (in Buber, 1975, 36–37).

But me, I am not the Baal Shem Tov. Actually, I am, as it were, an attendant lord; perhaps even one of his disciples. Lately, though, I think teachers are presumed by our society to be the Baal Shem Tov, and when disaster strikes, it is to the teachers that the political world turns paradoxically, as both cause and solution of present conditions.[1] That is, the sorry state of society is directly tied to the inadequacy of teachers who ought, it is scolded, to pay greater attention to standards, which will be set by political hacks, and repair the social fabric and maintain America's greatness. The decline in American power, the argument goes, is directly tied to the incompetency and inadequacy of teachers. Of course, as even a little consideration would indicate, there could be little correlation between America's power and its test scores, though remarkably, the conservative right continues to decry the public school system as the cause of America's woes. Gerald Bracey (2005) writes,

> If you think critically about this myth for a moment, the disconnect between test scores and a nation's economic health is obvious. Only the very, very foolish would think that 13-year-olds' skills at bubbling in answer sheets would mean much for a nation's well being. As if to prove that contention, the other "Asian Tiger" nations, also atop the world in test scores, saw their economies tanked during the mid-1990's. Singapore, which typically outscores *everyone*—it scored well above even other Asian nations in

all comparisons in TIMSS2003—declared itself in recession in 2001. The slump continues: In 2003 its per capita GDP was $23,700, down from $26,500 in 1999.

But in this heinous charge of inadequacy, every teacher is made culpable *and potentially treasonous* every time he or she walks into the classroom, and doesn't fulfill the expectations of some bureaucrat who ignorantly and irresponsibly levels this accusation. It would seem, to my mind, that there could be no more dangerous profession than teaching this day, though the rewards for assuming that risk are so remarkably low.

And what is this all about, Alfie? In my daughter's school, a student in her Advanced Placement (AP) American History class spouts malicious and derogatory talk about Jews and Blacks in his Chemistry class. I do not believe for the moment that his is an isolated case. And I am supposed to be concerned that his scores on the AP history test may portend America's economic decline! If this is what economic superiority requires, then I say, let's aspire for second best and focus on something really important, like the development in our children of an ethics of care and responsibility and for a Deweyan pragmatic immersion in the democratic beliefs and practices upon which the United States must rest. In the meantime, our children listen to our leaders regularly prevaricate, and our children daily bear witness to their unethical, even illegal behavior. Our children hear our legislators decry and declaim the horrors of education to which the children are subject, and yet on the very floors of the U.S. Congress uncivilly debate among themselves in public—in front of our children!—issues of life and death and war and peace in language and behavior I would hope no classroom teacher would tolerate, much less use! We would have, in these moments, I think, great need of the Baal Shem Tov, but alas, he has passed.

The story continues:

> The Baal Shem Tov died, and his place was taken by the Baal Shem Tov's disciple, Rabbi Dov Baer, the Maggid of Mezritch. When disaster threatened, the Maggid would go to the same place in the forest and he would say, "Master of the Universe, I do not know how to light the fire, but I can say the prayer. And that must suffice." And the Maggid would say the prayer, and the disaster would be averted.

I began teaching high school English in Sayville, Long Island in 1969. I taught ninth and tenth grades, and we read and talked about books and grammar. In 1971, I took a forced two-year hiatus from teaching; I started graduate school, and then worked with my father until we haply and, to my mind, at least, happily went out of business. In 1973, having spent several months gainfully unemployed, I returned to the profession as a temporary Reading Teacher at Eastern District High School in Brooklyn. Then, for the next fourteen years I taught English in the Great Neck Public School System on Long Island, New York. When I taught there, Great Neck North High School, though it has been originally erected in the 1920s, was yet a beautiful and stately structure, well befitting the lordly community over which it looked. Over the years the building had been well-maintained and regularly renovated. Indeed, when in 2005 I visited friends still living in the community, I returned to a high school that had once again been renewed, and on which a large addition had been recently built. Today several new gymnasiums, myriad new and renovated classrooms, sufficient teachers and supplies, and expansive grounds are part of the educational environment of the children who are fortunate enough to attend Great Neck North High School. The district even maintains its own computer repair facility, and of course, there is a sufficient supply of computers for everyone. When I taught in Great Neck, I complained regularly as I am wont to do, but those were wonderful years in the classrooms. I lived in the heady environment of academia—when what I did seemed important and crucial. Indeed, given the nature of the community and its expectations, what I did was important and crucial. Though now I often wonder what it is I thought I did. I had no idea how to light the fire, and I did not even know the appropriate prayer.

Great Neck is one of the highest socioeconomic areas in the country. Andrew Smith, in *Newsday*, writes that "Great Neck was prime real estate for centuries, long before theater stars, captains of industry and well-to-do commuting suburbanites made it the winter anchor of the Gold Coast" (http://www.newsday.com/community/guide/lihistory/ny-historytown). Great Neck is a community populated with highly educated people who occupy powerful and successful positions in business, the professions, and the arts. Now, the Great Neck District School District draws from a variety of incorporated and unincorporated locales of which Great Neck Village is itself only one. The 2000 census reports that the average family income in Great Neck Estates was $182,980.00 (http://www.newsday.com/business/realestate/ny-census-percapitaincome). In Kings Point the average family earnings was $229,423.00, and in Kensington the average was $156,311.00. In the Village of Great Neck, with a population of about 10,000, about 30 percent of the adult population have earned advanced academic degrees. That figure represents three times the national average. When I worked there, from 1974 to 1988, the student parking lots were forever filled with the newest and most stylish cars. The laboratories were all sufficiently stocked with the finest chemicals and newest equipments. There were colorful maps on the walls, and even the newest countries had representation. In the 1970s and early 1980s, when I served as teacher in the district, the business classrooms taught typewriting on an ample supply of new electric typewriters, and in the early 1980s, I clearly remember an in-service experience in which I was enrolled where I learned word processing on "the new computers." In that particular class I learned to physically move text around. I still love to do that, both literally and figuratively. The book rooms at North High School were filled from floor to ceiling with books. Real books. New books. Novels and trade books and textbooks. An English course on popular novels actually taught recent best sellers, and though over the years the number of electives has decreased, the quantity of books seems to always remain the same and current.

Teachers were fairly well paid in Great Neck, by teacher salary standards. The AFT reports that in 2004 the average teacher salary was $46,597 (http://www.aft.org/salary/2004/download/2004AFTSalarySurvey.pdf), but today in Great Neck, the average salary is $85,000.00, and about 24 percent of the teachers earn in excess of $100,000.00. And the work environment in Great Neck was, if not flawless, certainly without flaw. The rooms were plentiful, large, and clean, the hallways wide and well-lit. I remember that every evening the custodians would clean each and every room, put the chairs back in a preferred arrangement—I preferred circles, but my colleague liked rows. The custodians would wipe and wash the chalkboards clean in preparation for the next day's class.

When I worked there, Great Neck operated with an open campus—which means that when not in class, students could leave the building and, in warmer weathers, sit on the expansive front lawn. Students were free to walk the building or go to the library to read the abundant supply of the latest newspapers and periodicals, or to browse amidst the stacks and stacks of books. Seniors were permitted to drive home or into town for lunch. Within the main school building itself, an alternative educational experience was available—the Community School—for students who desired a more intimate and personalized education in very close proximity to the traditional educational system. At the South High School a similar minischool existed called SWAS—School within a School. Generally, there were between forty and fifty students enrolled in these constructs, and it maintained its own core of dedicated faculty. For years students enrolled here were not obliged to come upstairs for any of their academic work, and indeed, often took lunch within their separate facilities as well. They lived, as it were, by a separate daily code of educational conduct. Mary Ann Raywid said somewhere that a school with over 400 students ought to be broken up into smaller community schools, and Great Neck had put her desire into practice.

In fact, during my final four years in Great Neck I was a teacher at the Village School, a public alternative high school that functioned as an integral part of the Great Neck School System. Originally begun in the heydays of experimentation and rebellion during the late 1960s and early 1970s as a truly alternative educational experience, the Village School became in the 1980s a repository for the disaffected and disgusted, for those who today might be referred to as the emotionally challenged. There was usually thirty to forty students enrolled in the Village School at any one moment, and when I worked there, it was staffed by five teachers and a full-time psychologist. There was a great deal of freedom in this school: we taught what we wanted and what students desired, and we were not obliged to take any of the burgeoning array of state tests that New York began to mandate with ever-increasing stridency. When I taught at the Village School, in addition to the standard English fare, I also was able to teach history, film, and during one semester, I sponsored a Health Class. In this experience, once each week, students and I would shop in the local vegetable stores and prepare a lunch in the school kitchen, sit and eat that lunch in the dining area. We would discuss nutrition and our lives. One year at the local Goodwill I purchased my own classroom furniture from school funds, and we always bought whatever books and supplies we needed, usually from the local bookstores. We maintained our own building separate from the main schools. And in Great Neck, besides these two secondary schools in which I worked, there existed Great Neck South High School serving the other end of town, a middle school for each side of the community, and eight or nine neighborhood elementary schools.

Great Neck was a wonderful place to teach. Well, of course. Most of the students went to college—the community was wealthy enough to buy everyone a college education if they so desired. I think almost 90 percent of the graduates went on for higher education. When homework assignments were made, then assignments were mostly completed. Books were read and studied; math and science courses were advanced and filled. Supplies were not unending, but were certainly plentiful.

As the Maggid of Mezritch, I seemed to be in the right place of the forest and I knew how to pray, but I had forgotten how to build the fire. I was too far from wisdom, but I felt a passion.

Why do I tell this story now?

I am a university professor of education. I prepare preservice students who desire to be teachers, and I teach practicing professionals who continue their own studies of curriculum and instruction in the graduate school. I often wonder what it is I am supposed to be accomplishing. I tell my students that teaching is a noble profession, despite its low status in society. I tell my students that in order to be a teacher one must first be very smart. I try to help my students learn to be very smart. In the clichéd world of the school, I want my students to become lifelong learners, whatever that empty phrase means. I want them to continue to read and study texts, to understand the complex social web spun that we term education. In Tillie Olsen's (1960) words, I want them to know that they are more than "this dress on the ironing board, helpless before the iron" (12). Or to offer my own arachnoid metaphor, they are more than food trapped in the web by the spider for later leisurely repasts. I teach them the place, and I teach them not the prayer, but how to pray.

Ah, but I cannot teach them to build the fire. I have forgotten how to do that myself. And what then must I do?

Now, in her junior year, my daughter is suffering through chemistry. I make less money than her chemistry teacher. But she is taught by what I hope and expect to be a competent chemistry teacher. My daughter does her homework, and completes her labs, and struggles desperately with the math of chemistry. In my household, math is not a strong suit—I suffered myself through mathematics in high school and college, and the last formal math course I took was symbolic logic in 1969; I chose it because it was a math course with a minimum of numbers. Though I intended originally to study medicine, I couldn't manage the math or chemistry courses that were required, and when they told me I also had to take physics, I switched majors. And so for our daughter, we hire tutors from the university and pay them highly for their work.[2] Slowly, I think, she is learning and gaining confidence in the subject matter. Of course, Emma wants to go to college, and so she frets about her grade point average, and I commiserate with her concerning the harsh effects on her GPA of her present chemistry and math scores. I assure her that her grade will improve as the semester continues, and I anguish over those grades that are horribly misshapen snapshots of her real effort. I do not think my daughter is alone in her feelings, but she is my daughter, and I know her.

I look about for the Maggid of Mezritch and for what will suffice, but alas, he has passed.

> When Dov Baer, the Maggid of Mezritch, died Moshe Leib of Sasov took his place, and when disaster threatened the Jewish people, Moshe would go to the same place in the forest and say, "Lord of the world, I do not know how to light the fire nor how to say the prayer, but I do know the place and that will have to suffice." And always it did.

I do not think that anything occurred at Great Neck North High School in New York that is not still occurring at Emma's school here in Menomonie, Wisconsin. There and here teachers teach. The myth that

American schools are plagued by incompetent teachers is nothing more than that: a myth. Berliner and Biddle (1995, l08) show unequivocally (that means with statistics!) that "the teaching force of America does *not* consist of bottom-of-the-barrel people who are deficient in intellectual ability and poorly trained by their colleges and universities. Despite the enormous size of the teaching force, evidence suggests that the average teacher in American is talented, high achieving, and well educated" (italics in original). That is, in 1995 as in 1985 and in 1975 and in 1965 (when, alas, I graduated Jericho High School), teachers taught with skill and commitment and subject knowledge. And the students in Great Neck North High School (as in Jericho High School in the class of 1965) learn as do students today at Great Neck North High School and Jericho High School and Menomonie High School . And as they do in many, many other schools across the United States. But, in Great Neck and in Jericho, the environment supports not only the learning, but also the college expectations and realities that stem from that learning. In Great Neck, the resources are available to the schools *because* education is so highly valued, and in Great Neck resources are available to the school *because* the community has the resources to offer to them! In Great Neck, what teachers do *seems* important because the community *deems* it important. And that community, therefore, commits the ample resources it has available to fulfil its educational *and* social responsibilities.

In Menomonie, not only are the values concerning education different than they are in Great Neck, but the resources available here, too, are different. Though the school teachers here receive middle-class salaries, these salaries represent only half of what Great Neck teachers earn. Yet the subjects they teach are pretty much the same. Chemistry is yet chemistry, and despite the (modest) range of intellectual achievement of teachers, and despite the differing ideological situations toward their subject matter (intelligent design versus evolution; string theory or particle physics, *Romeo and Juliet* or *Julius Caesar*), the courses are, in the main, comparable. Furthermore, because it is true that in Menomonie employment is not perceived as driven by a desire for an acquisition of higher education, the percentage of students who go on to higher education is about half of what is in Great Neck. In the latter, a higher education is crucial for the work places from which these students have the opportunity to choose, but in Menomonie, as in so many communities in the United States, those opportunities are severely limited and increasingly narrowing. Indeed, Berliner and Biddle (1995) suggest that "Americans have overstressed the tie between qualifications and employment" (101). In America, too many college graduates are underemployed; that is, they are working in professions that do *not* require the intelligences that they have developed in colleges and universities. Indeed, our economy cannot sustain the highly educated. Higher education has nothing to do, in this case, with satisfactory employment. Nonetheless, in this home at least, college has always seemed a given, and so Emma's learning in chemistry and mathematics becomes an issue, whereas in other homes it would not be treated with the same concern as it is here.

I am certain that the emphasis on accountabilities, exacerbated now by the heinous No Child Left Behind Law, has created a situation untenable in many schools. This focus on accountability, which assumes a level of incompetence in teachers and which conflates the acquisition of an education with the productive efforts of a worker on a factory line, requires constant observation and standardized measurement; the levels of fear and stress this situation produces in my child, at least, concerns me. The testing regimen at Menomonie High School appears to be dangerously oppressive. Every single day my daughter seems to have a test in one of her courses. I am wondering when she has the time to be taught! The school seems to hold to the idea that there is nothing in a child's life except schooling.

I have thought about this for all of the years I have spent in the class-rooms. I have wondered about what is going on in our schools, and what it is I think ought to be going on there. I mean, we know the place, but I am certain we have forgotten how to light the fire and to say the prayer. How else to explain the vicious and contentious battle over education in the United States? But remarkably, what I have imagined is going on and what actually occurs is remarkably similar. For the most part, there is sub-ject matter that is to be transmitted, and in the majority of schools across the United States, that content is the curriculum. As you might suspect, I am not enamored of the American school curriculum; as a curriculum theorist I despair over what is called education in American schools, and I am often appalled by what transpires in the name of education. However, in the good schools and in the bad, this is what occurs: teachers teach, and for the most part, students learn. In school districts like Great Neck, stu-dents learn not because they are inherently more intelligent, but because their learning is supported everywhere in their lives: at home, in school, in the community. In Great Neck, that learning is *also* supported by remark-able financial resources, not unlike those that supported the elite class in the nineteenth century who attended private schools and who would serve as the power elite. In Great Neck, that learning *seems* to have real consequences. In places like Menomonie, learning occurs as well, though with less intensity and less meaning. In too many schools, learning is seri-ously obstructed by conditions completely outside of the control of teach-ers and students. Jonathan Kozol (1992, 2005) writes passionately about these situations about which our American society should be ashamed. We teachers know, however, the place, though how to light the fire and the prayer to recite have been lost. We move forever and inevitably further from the Ur source.

What do *I* then tell my students? I tell them that in the best of public schools and in the worst of them, in the best of times and in the worst, there remains a certain consistency. We will in our university class explore curriculum theory, but they must learn to settle in their professional lives for curriculum development and classroom management. They each have their own individual educational tales of woe, and though they aspire to turn these horror tales into fairy tales, alas, there will be precious little opportunity for them to do so. We have forgotten the prayer. Though we may despair at the continued imposition of external standards and means of evaluations, and complain at the continuing erosion of our autonomies and confidences, there are severe limits to our power. The fire is lost to us. We would change the world teacher by teacher and child by child, but alas, we are too philosophically diverse, and besides, we haven't yet had our morning coffees.

The grumbling has always existed. No sooner did the Israelites leave Egypt than did they begin to complain. Did God take us out of Egypt, they whined, only to have us massacred before the Sea of Reeds, or worse, to drown in the Sea escaping from the hordes of Egyptian soldiers following us to return us to slavery? At least in Egypt, they claimed, there was food and water! At least under slavery, they whimpered, we had a roof over our heads. In Egypt there was repast beyond manna. The Israelites were, as perhaps we are all, a stiff-necked people. Complaining seems endemic to human nature—we are forever unhappy with whatever state we are in: we could always be happier, I think. And the schools about which we complain have been the schools about which we always have complained. We know the place. We complain on the one hand because our work requires an effort we did not fully anticipate, and as a result, we are forced out of a certain complacency and easiness. And we complain because we know that though the schools are in need of improvement, sometimes dramatic overhauls, it is not the schools that are to blame for the social malaise for which the schools are blamed. Despite the vicious rhetoric aimed at our competence, we are not incompetent.

Rabbi Israel of Rizhyn became the leader when the Maggid of Mezritch died, and when disaster threatened the Jewish people, Rabbi Israel would say, "God, I no longer know the place, nor how to light the fire, nor how even to say the prayer. But I can tell the story, and that must suffice." And it always did.

Open school night in Great Neck was always a well-attended and complete event. Classrooms were packed filled with parents anxious to learn what the school program would be for their children. Students sold refreshments they had baked at home for the many extracurricular events that required funding—the Air Force still usurped most of the funds that would have subsidized these activities—and administrators walked the halls greeting parents and ballyhooing the district. I waited in my classroom for the parents of my third period class. Mr. Walters (a pseudonym, of course) entered my room smiling, and walked right up to me. He introduced himself, and he asked, "So, how do you find students these days?" I knew to what he referred—the myth that the quality of students had been in serious decline since Mr. Walters had graduated high school. I looked at Mr. Walters and I smiled. "I don't have to find them at all, sir. I come to my room when the bell rings and there they are, in their seats!" Yes, in most schools, the students are always there in their seats. Waiting. And usually hoping. We are supposed to teach them biology, history, chemistry, mathematics, and the values of great literature. In Great Neck. In Menomonie. In Brooklyn. In Topeka. In Grand Forks. In San Francisco. In Atlanta. In Tallahassee.

But I think we have forgotten the place, and the prayer and even how to light the fire. No, we are not in any sense a species degraded; we are just fortunately farther from Truth than too many of us would readily admit. If there were ever a single place where the truth might settle, we no longer know where that place might be. And if there was any supplication that might plead for redemption, we no longer remember it. Thank goodness for that, I say; I am weary of our Days of Prayer and failures to act. And that fire—what was it for actually, that ritual act, which only in the right place and with the right prayer would fulfill the conditions for divine intervention? Other fires were daily built, and other prayers were daily recited, but none of them ever brought divine relief unless performed by the saintly Baal Shem Tov. And though the conservative fundamentalist right insists they are the possessors of truth, others of us are more humble; we claim to have learned and to not know.

What we have left then, is the story. The story is not the truth; rather, the story is our particular narration of the world in which we all mean to live. And the story occurs in the space created between hope and memory. Levinas (1990, 66) writes, "Isn't the future world the possibility of rediscovering the first meaning, which would be the ultimate meaning, of every word?" But we do not live in this future world; we live in the present, and we work toward the future. We search for the ultimate meaning and are relievedly frustrated in that quest. You see, if all we had was Eden, there would be no need for hope; everything would be already there and immediately present. We left Eden, ah, by intelligent design: we ate know-ingly and willfully of the Tree of Knowledge. And now, we have to work for our bread. We struggle because outside of Eden the world is not perfect, and we must make it so. But as John Lennon said, "It don't [*sic*] come easy." Thank goodness for that. And Lennon also urged us to imagine. We hope for better, and we remember what we thought we once had; we imagine having it again. Out of this dialectic comes the story: in the space between memory and hope. But if we ever knew the place, we have lost its location; and if we ever knew the prayer, we have forgotten it; and if we once knew how to light the fire, well, we no longer possess that skill. We have only the story, and that must suffice.

Curriculum is the story we tell our children; it arises in the space between hope and memory. Sometimes, the story concerns Columbus, and sometimes it concerns Holden Caulfield. Sometimes the story is about the story of Columbus, and sometimes the story concerns the importance of trigonometry. Sometimes the story explores the formulas for gravitational force, and sometimes the story concerns the necessity and power of raising one's hand in class before speaking. Sometimes the story concerns anoth-er's sense of our own worth, and sometimes the story creates that worth. *But it is all story.* And the great story tellers well, they know how to tell stories to instruct, to entertain, and to enthrall the audience, and to use the space between hope and memory. The great story tellers know that behind the important stories are ethics and relationship, and not truth. The great stories will suffice to save us.

Story and Being

What connects the teachers in Great Neck to those of Menomonie and to those in the South Bronx or any other place in the United States is that each stands in the classroom not as the bearer of truth, but as the possessor of wisdom. Teachers are the tellers of stories. We anticipate our future and bear witnesses to the past. We teach the story, and in the telling of stories, we teach storytelling. We come to being in the telling of stories.

The story constitutes our self. Oliver Sacks (1985), the neuropsychol-ogist whose work studied the effects on being of the loss of the ability to create story (105–106) concludes: "[E]ach of us *is* a biography, a story. Each of us *is* a singular narrative, which is constructed, continually, unconsciously, by, through, and in us—through our perceptions, our feel-ings, our thoughts, our actions; and not least, our discourse, our spoken narrations...To be ourselves we must *have* ourselves—possess, if need be re-possess, our life-stories. We must 'recollect' ourselves, recollect the inner drama, the narrative, of ourselves." This drama and this narrative arises out of the fact that we live with others for whom we are responsible. In the acknowledgment and acceptance of that mutual responsibility, we come to exist. Ira Stone, in his study of the philosophical relevance to Western life of Levinas' (1998, 8) Talmudic discourses says, "We are created by another's love and create ourselves by accepting the burden of this love and its obligations." We are created in the story which tells of our life together, even as that story binds us to that life and that community. Levinas (1993) states that "Misery and poverty are not properties of the Other, but the modes of his or her appearing to me, way of concerning me, and mode of proximity" (18). How the other appears to me, her way of concerning me and his mode of proximity, comprise the story I tell of the Other and myself. The stories we tell are constitutive of our selves and our relationships.

The story arises out of the space between hope and memory. It anticipates a future that must arise out of the past. I can think of no other reason to establish schools for our selves and our children except as means to "gather up the dispersion of the sacred in the profane" (Levinas, 1994, 7). As our goal must be to be holy (*Leviticus*, 19:2), that goal can only be achieved through action. We will do and we will hear. Our stories are about that action in community and our stance toward the other. Levinas (1994, 131), who insisted that the wisdom of the Hebrew be added to that of the Greek, writes that Judaism is "preoccupied with behavior, action and rites, the practical orientation which consists in reading the Bible through the Talmud (the relation with God always being mediated by one's neighbor and by the defense of 'the poor, the widow, the orphan, the stranger') . . ." For Judaism, Levinas writes, the story constitutes and concerns our ethical beings; the existence of story, lies, perhaps, at the heart of Judaism. As our purpose in the schools is not dissimilar, so should the stories we tell constitute the central matter of the classroom, and ought concern our ethical being. These stories are what must suffice.

The Story and the Second Set of Tablets

I want to situate the origin of story in an event that derives from a culture long excluded from educational discourses. Let me consider that the impetus for stories arose as a result of the existence of that second set of Sinaitic tablets. Perhaps at this foundational moment of Judaism we may discover the centrality of the story. Moses, you will recall, broke the first set of tablets that God had carved and on which he had written the commandments.

But Moses smashed those tablets when he beheld the people dancing about the Golden Calf. Then, God told Moses to return to the mountain's top with a second set of tablets God had commanded Moses to carve himself, and on which God would again write the commandments: "Carve for yourself two stone tablets like the first ones, and I shall inscribe on the tablets the words that were on the first tablets, which you shattered" (*Exodus*, 34:1). On that first set of tablets were written God's commandments that the people promised to obey and to hear. But before they had even heard them, they had made the Golden Calf. Perhaps these commandments were too exalted and too far removed from the exigencies of daily life, perhaps, not unlike the homework assignments that remain undone and not comprehended. There was no story in them, only command.

Perhaps, like the place, the fire and the prayer of the Baal Shem Tov, the first set of tablets existed "for themselves"; they were too singular. In the absence of any dialogic relationship, of any sense of covenant, the commandments were authoritative, and rendered human kind abject: stories would not exist. The Israelites had escaped slavery; they were not keen to return to a different form of it. Levinas scholar and translator, Richard Cohen (in Levinas, 1999, 7) writes, "Those first tablets were *too* first...too exalted, cut too far above the measure of humanity...Too pure a divinity yields only a debased humanity. The high and the low would have to be brought into conjunction, into unequal reciprocity; *religion* (*re-ligio*) would henceforth be necessary for the very humanity of the human and the divinity of the divine." In that bridging between the divine and the human, God "transcend[s] God's own transcendence" and come down to humankind. As God cares for the poor, the widow, orphan, and stranger, so, too, should we care for the less fortunate among us. "Biblical humanism," Cohen writes, "requires the meeting of divine and human through moral command manifest in moral action, hence in human courts, divine command through human response, human response divinely commanded, absence and presence, preoriginality and originality" (in Levinas, 1999, 11). That meeting between God and humanity occurs, I will suggest, as a result of the second set of tablets.

These second set of tablets are a reminder of the first: but their existence established a relationship between humanity and God and made possible the story. In his introduction to Levinas' *New Talmudic Readings*, Cohen (1999, 9) notes of this second set, that "coming second, taking second place, the other-before-the-self, the other-put-into-the-same, converts firstness—the for-oneself, *of both God and humanity*—into an actual *relationship*, the most concrete and pressing of relationships, the exigencies of morality and justice, breaking the bounds of both terms." The necessity and reality of that second set of tablets established a relationship. After God writes the words on the second set of tablets, "God said to Moses, 'Write these words for yourself, for according to these words have I sealed a covenant with you and Israel'" (*Exodus,* 34:27). Those words are the stories, and if we understand "according to" to mean "bound by," then the second set of tablets commits *both* humankind and God to a relationship founded on ethics and justice. In the story that derives from the second set of tablets, "Moses received the Torah from Sinai, and transmitted it to Joshua; Joshua to the Elders; the Elders to the Prophets; and the Prophets transmitted it to the Men of the great Assembly. They said three things: Be deliberate in judgement; develop many disciples; and make a fence for the Torah" (*Pirke Avot,* 1). Once written, the story required telling and exegesis; that relationship produced the story and maintains storytelling. The story lies between hope and memory: we hope to be holy like God, but we remember what a stiff-necked people we are. The second set of tablets established the necessity of the story by establishing in the covenant a relationship. The covenant is about relationships of ethics and justice. The stories remind us of our glorious and inglorious past, even as they portend a future. This process ought to be the substance of our classrooms: the matter of memory and hope. Teachers have only the stories to tell.

We become teachers when we tell the stories to our students, and we come to exist as teachers at least in and by this dialogic relationship. When we accept the other, our students, by and in the telling of stories, we ourselves come into being. What we do in the classroom is to tell stories. It is our struggle in life to understand, to acknowledge, and to live by the stories spawned by the existence of that second set of tablets. Our libraries and school buildings were constructed to house the stories derived from the presence of that second set of tablets, even as the Tabernacle was created to house these tablets that would serve to remind us of that first broken set. The second set of tablets—the same but different, the other in the self— reminds us of the standard to which we aspire but cannot achieve: this we will do, ah, but fall forever short of doing.

The second tablets require us to tell and to listen to the stories in order to discover what we must do. It is what Jews have been doing for millennia. Levinas argues that this understanding is best done *not* through criticism but through exegesis. "Exegesis...is text interpretation not through *explanation* derived from *objective* context alone, but through *understanding* derived from the text's as well as the subject's own *subjective* context" (in Levinas, 1999, 22). Exegesis acknowledges our being in the context of our relationships. Exegesis is the telling of stories. When God threatens to destroy the Israelites in the wilderness for their complaining, Moses, like Abraham, negotiates with God to save the people: "Why should Egypt say...'With evil intent did God take them out, to kill them in the mountains and to annihilate them from the face of the earth?' Remember from Your flaring anger and reconsider regarding the evil against Your people. Remember for the sake of Abraham, Isaac, and Israel..." That is, Moses says to God, remember now the story, and be merciful. It is when God relents that God commands Moses to return to the mountain top to acquire the new tablets and to write down the words. Those words are the stories we tell. They concern how to become a holy people, a nation of priests. Ironically, our stories in the classroom will achieve what the conservative right would have for our schools and our children. The stories are all that remain. It must suffice.

Chapter 6

Study and Benevolence

I am finishing my sixth decade. Every third thought is of the brave.

For thirty-five years, I have been a teacher. I think continually of teachers and the work that we do. We stand not only vulnerable in front of hundreds of students daily, but also stand defenseless before the public eye, serving as easy prey to politicians and government bureaucrats whose motives are, at best, not pure. William Pinar (2004, 30) writes, "We teachers are conceived by others, by the expectations and fantasies of our students and by the demands of parents, administrators, policymakers, and politicians..." Subject to the contumelies of everyone who knows better, teachers are regularly castigated in public for too many real and illusory social ills and national failures for which teachers could not possibly be responsible but for which nevertheless, teachers are made to bear responsibility. There seems to be no end to the accusations maligning teachers and their work, no criticism too harsh to aim at the schools and their staff, and yet daily it is off to work we go. I am finishing my sixth decade. Every third thought is of the brave. I have, after all, been a teacher for thirty-five years. I know what they say about me, but I take the time now to consider what it is that I have been doing, and that, I suspect, I will continue to do for the next few years.

The question concerning what teachers do is neither a casual nor an obvious one. Of course, teachers teach, but what and how they teach remains central to the development of the individual and society, and as a result of this effect, always subject to ideological contention. What exactly it means "to teach" has never received definitive designation, nor has what it means to learn been unanimously agreed upon. Nor has what it is that must be learned been definitively determined, and yet is forever and inevitably subject to every political wind. Though John Dewey (1966,

51) long ago wrote that "education means the enterprise of supplying the conditions which insure growth, or adequacy of life, irrespective of age," it has never been clear of what that growth, adequacy, or life should consist. Finally, though the rhetoric of most educational mission statements claims that life-long learning is its goal, and though theoretically, education remains ideally appropriate for all, formal schooling is reserved for the precious few and relatively young, and increasingly, of late, the more financially privileged.

Nevertheless, the growth to which Dewey refers is both personal and social: individuals, after all, comprise the social, and because the social is greater than the sum of its individuals and completely nonexistent without them, then the social must be taken into consideration and elaborated upon in the construction of the educational processes that, for the most part, take place in society's schools. For at least our children and not a few young adults, it is the school that society officially designates with express educational purpose. And if society sends people there, that is, to school, then society expects that something happens there to them, at school, and it is the teacher who is assigned the responsibility for facilitating that happening.

Every third thought is of the brave.

It is no small act to engage—or attempt to engage—daily with students who in the best of circumstances desire a knowledge they cannot yet imagine, and for which the teacher, a stance I have taken over the past thirty-five years, is responsible for not only the communication of that knowledge but its reception as well. It is a brave act to engage daily with those who in the more common circumstances, sit complacently and too politely awaiting the clang of the next bell, passively resisting much of what I say; it is a brave act to engage daily with those who in the worst of circumstances actively resist my offer and refuse my presence. At every minute of every day, by her very presence the teacher reminds the students that they are not yet perfect, that the achievement of that perfection is a goal toward which all must be directed though it can never be achieved, and that the effort that must be made in the seeking is arduous and difficult. It is a brave act to speak of hope to those who live without it.

I am finishing my sixth decade. Every third thought is of the brave.

Robert Frost's observation that "home is where when you go there, they have to take you in" remains true as well for the school: school is where they take you in when (or because) there is no other place for you. Whether the intent be social, legal, or educational, society designates that the school serve as the harbor before life's great sail, and that all below a certain age must take up regular occupation there. What must be done at that harbor stands at the center of the struggle for the American curriculum and the battles over school control. Whether the time at harbor should be spent thinking about the activity of sailing, or whether it should prepare us generally (or specifically) for the accomplishment of the eventual sail itself, or whether and how time spent at harbor might designate and prepare each individual for a particular role in the ultimate sail remains part of the contentious conversation concerning education in the United States. We debate how to prepare for the inevitable sailing while yet remaining on dry land; but since we too often doubt the relevance of preparing for the eventual sail when we are not actually yet sailing, we explore the height of the mast as the main topic. We wonder in the schools how (and why) to build boats and sails when there is, or might be, no legitimate need for them? We wonder if sailing is all that is to be taught while remaining yet at harbor. In the schools we ponder about what must be accomplished when students might be out at sea, but are in the present not sailing. We vigorously debate what matters must be the subject of teaching, and too readily remain deaf to the voice of the professional who must accomplish the impossible—the teacher. At every moment of every day the teacher addresses questions large and small, and must attend to her effort despite the difficulties or unwelcome natures of the task or object, and the demands and resistance of the students.

William James (1892/1961) held that the hero is one who makes the effort, which for James seems to be equivalent to the acts of attending, willing or consenting. The hero is one who unwaveringly sustains the idea regardless of the myriad tendencies to let it slip away. "*To sustain a representation, to think*, is in short, the only moral act, for the impulsive, and the obstructed, for sane and lunatics alike" (320, italics in original). It is, I think, the sustenance of the idea that leads to action. The alternative to this effort is passivity: the incapacity to sustain an idea is to lack will, incomplete and unfinished activity, and to be incapable of choice and remain, therefore, unwilling to act. In such a situation, we lie abed unable to rise, or we await someone else's command and direction. In *Psychology: The Briefer Course,* William James (327) writes that "...not only our morality, but our religion, so far as the latter is deliberate, depend on the effort we can make. '*Will you or won't you have it so?*' is the most probing question we are ever asked; we are asked it every hour of the day, and about the largest as well as the smallest, the most theoretical as well as the most practical, things...We answer by consents or non-consents and not by words...what wonder if the effort demanded by them be the measure of our worth as men!" As teachers, we answer daily with our consents and our nonconsents *and not only with our words!* I think we teachers must seriously consider what it is to which we consent, and to that which we teachers do not consent. We must wonder what it is we teachers actually do and say.

Teachers are heroes because at every moment of every day, teachers consent or refuse to consent. At every moment of every day, school depends on the effort teachers make. Or rather, at every moment of every day, the type of classroom that students experience depends on the effort of the teacher. In the daily functioning of their professional lives, James (1962, 95) urges teachers to "See to it now, I beg you, that you make freemen of your pupils by habituating them to act, wherever possible, under the notion of a good. Get them habitually to tell the truth, not so much through showing them the wickedness of lying as by arousing their enthusiasm for honor and veracity. Wean them from their native cruelty by imparting to them some of your own positive sympathy with an animal's inner springs of joy." James knew how difficult it is to be a hero, and how resistant society could be to the hero's will. And James knew that some of those Napoleons out there might really be so: indeed, it was their capacity to sustain the idea that led the rest of us to entomb them in the asylum. Though we treat our prophets as lunatics, some of our lunatics are, indeed, prophets. Well, yes, and some are, indeed, lunatics. But it is often history that serves as the arbiter and declares whether the sustained idea is that of the prophet or the lunatic. In this life, we must yet act without history's judgment. Here, in this life, the moral act is to sustain the idea—the exercise of will and effort. And this is the only moral act because it is only this that leads to action. I would add to James' idea only this: that all acts must be founded first on an ethics, and I suggest that the ethics of Isaiah seems relevant here for the teacher. Isaiah admonishes the people for repenting in form but not in substance.

> Can such be my chosen fast, the day of man's self-denial?
> To bow down his head like a bulrush, to sit in sackcloth and ashes?
> Is that what you call fasting, a day acceptable to the Lord?
> Surely, this is the fast I choose:
> To break open the shackles of wickedness,
> To undo the bonds of injustice, and to let the oppressed go free, and
>   annul all perversion
> Surely you shall break your bread for the hungry, and
>   bring the moaning poor [to your] home;
> When you see a naked person, clothe him; and never turn from
>   your fellow.   (Isaiah 57:5–8)

Perhaps the teacher's sustenance of *this* idea would, indeed, create an education that would surely change the world. And would create a world changed by education.

On December 24, 2006, I attended a party billed as a "Christmas Eve Party for Jews." I think I remember that the actual name of the party was something to the effect of "Hey, he was a Jew, Anyway." It was a lovely party: the food and drink were delightful, the home elegant and comfortable, and the hosts and guests amiable, voluble, and articulate. There were plentiful teachers in attendance. One conversation in which I engaged with considerable interest developed over the use of the term "in the trenches" when referring to the situation of teachers in the schools. The man with whom I was speaking mentioned that he worked in the schools as "an administrator" and not "down there in the trenches." I said, too apologetically, I think, that teachers often use the term in the trenches with a certain degree of pride; in my years teaching in the public high school, I had myself used the term with some vainglorious satisfaction, but I had since learned to reject the place of that vanity. The trenches is not a fair or helpful place in which to reside or from which to speak, though it is true that to be in the trenches has come to be viewed as superior to being, well, in the offices, in the administration buildings, or in the ivory-towered universities. From their self-designated position "in the trenches," teachers proudly function under deadly fire. In the trenches describes the environment in which teachers boast that they work. I worked daily as a teacher.

I responded to my associate that I could not fathom why teachers would want to willfully consider themselves as somehow superior for adopting this position relative to all other school personnel and policymakers, including politicians. I said that teachers wore this description of working in the trenches as a sign of some moral superiority, but that it seemed to me that the concept of being "in the trenches" demonstrated a serious misrepresentation and a dangerous denigration of the teacher and his or her work. Such a designation positions the teacher suffering defensively in a veritable war zone, working in the midst of mortal assault, only minimally in control over his or her actions and in a position to be only minimally aware of events. After all, from the trenches what may be seen is severely limited. I objected strongly to the metaphor. I still object.

To be in the trenches refers to a position taken in war. I reject the idea that we teachers exist in such a lethal combat zone, with bombs and bullets whizzing over our heads as we huddle down safely out of range attempting merely to stay alive. Nor do I want to consider that from these trenches we periodically raise our heads (and weapons—ah, and what would those be, texts and lectures and learnings?) to fire back ammunition at the enemy in our attempt to defeat them. And would those enemies be the students in our classes, or the world in which the school and our classrooms are nested? I think that in these times one can get arrested for such aggressive behaviors. The image in the trenches suggests that the occupants of the school exist in a precarious and hostile state, and positions the teacher as engaged in deadly battle. Who would such fardels bear?

Indeed, if the country needed as William James (http://www.constitution.org/wj/meow.htm) argued, a "moral equivalent of war," that position ought to be not one where teachers hide our heads, but rather, one where they stand tall and proud in their disciplined position. From there we might argue that as in war the good of the whole supersedes that of the individual, then in the exercise of education we might together gather our resources to create "a moral equivalent of war" so that our schooled citizenry might engage together in creation of a "permanently successful peace-economy." James said hopefully, "In the more or less socialistic future toward which mankind seems drifting we must still subject ourselves collectively to those severities which answer to our real position upon this partly hospitable globe. We must make new energies and hardihoods continue the manliness to which the military mind so faithfully clings." James argued that in this moral equivalent of war, we might conscript our youth "to have the childishness knocked out of them, and to come back into society with healthier sympathies and sobered ideas. They would have paid their blood-tax, done their own part in the immemorial human warfare against nature..." This position, however, need not be defensive nor frightening, but challenging and exhilarating. If war is the victory of humans over other humans for generally suspect motives, then this moral equivalent of war should achieve the victory that war achieves, but it should triumph over nature—the unequal, uneven, and sometimes unnatural conditions in which humans live. "We should get toughness without callousness, authority with as little criminal cruelty as possible, and painful work done cheerily because the duty is temporary, and threatens not, as now, to degrade the whole remainder of one's life." Rather, this position would ennoble life.

I offered to my conversant the idea that teachers ought to, rather, assume the moral *high* ground; I urged that what the teacher does is *to rise* ethically in the world, and urges students to rise with him or her. Hardly remaining concealed and frightened down in the trenches, the teacher must stand proudly as model for the degraded world. The teacher should think of herself as assuming the stance of the prophet in the midst of a corrupt and corrupting society, and serve as its harsh critic and moral exemplar. Yes, indeed, I said to him, the teacher must appear *as a prophet*. Abraham Joshua Heschel (1962, 16) says, "The prophet's word is a scream in the night. While the world is at ease and asleep, the prophet feels the blast from heaven." This, I added, rather than that image of the teacher in the trenches crouched defensive and besieged, is a metaphor fitting to the position of the teacher: the moral exemplar for the degraded society. In this view, the classroom is the locus of ethics and life. Here, in that classroom, let the teacher stand tall and create the seeds for a better world. Not a pleasant position, but these fardels we have no choice but to bear. To be teacher is to be prophet. Yes, indeed.

What the prophet speaks is critique, and the words are, yes, harsh. To be a prophet is to be brave, even to be threatened, but to be a prophet is not to describe one's position as in the trenches. "[T]o be a prophet," Heschel (18) says, "means to challenge and to defy and to cast out fear." Such a position does not come without calumny or risk. Jeremiah, for example, complains,

> I have become a laughingstock all the days;
> Everyone mocks me.
> For as often as I speak I have to cry out,
> Have to complain of violence and abuse,
> For the word of the Lord has become for me
> A reproach and derision all day long.   (20:7–8)

Like so many teachers today, Jeremiah is rejected by his people, beaten and cast into prison for the accusations he hurls at the populace.

> I was like a gentle lamb
> Led to the slaughter.
> I did not know it was against me
> They devised schemes, saying:
> Let us destroy the tree with its fruit,
> Let us cut him off from the land of the living,
> That his name be remembered no more.   (11:19–20)

Like so many teachers today, Jeremiah harshly suffers, and in the exercise of his prophecy experiences what we now would describe as severe depression and suicidal tendencies.

> Cursed be the day
> On which I was born!
> The day when my mother bore be,
> Let it not be blessed!
> Cursed be the man
> Who brought the news to my father,
> A son is born to you,
> Making him very glad...
> Because He did not kill me in the womb.
> So my mother would have been my grave,
> And her womb for every great.
> Why did I come forth from the womb
> To see toil and sorrow,
> And spend my days in shame?
> O that I had I the desert

A wayfarers' lodging place,
That I might leave my people
And go away from them!   (20:14–15, 17–18)

Alas, too many teachers have echoed Jeremiah's plaint, but it is never from the trenches that Jeremiah speaks; rather, the prophet thunders his teachings from higher ground. His is not an easy, carefree, and selfish life, and yet, Jeremiah, having received the word of God, has no choice but to decry the degraded state of the people and urge them to return to the high moral plane set for them in *Leviticus*: "I am holy, therefore you should be holy." The prophet, not at all maneuvering unseen in the trenches, arching mortars and spraying bullets at the enemy from protective concealment, stands fully visible on solidly higher ground urging the repair of the world, and hopefully preparing its citizens for its repair. The Prophet Micah (6:6–8) says, "You ask 'With what shall I approach God, [how should] I humble myself before God? Shall I approach God with burnt offerings, with calves a year old? Would God be pleased with thousands of rams, with myriads of streams of oil? Shall I give my first-born for my transgression, the fruit of my body?' God has told you, O man, what is good, and what God requires of you: Only to do justice, And to love goodness, And to walk modestly with your God." I think that it is the bravery of the teacher that offers this instruction. The teacher as prophet does not operate from the trenches nor remains unseen.

Rabbi Hanokh (Buber, 1975, 314) told this story: Once there was a man who was very stupid. Every morning when he arose it was so hard for him to find his clothes that at night he almost hesitated to go to bed for thinking of the trouble he would have on waking to find his clothes. One evening he finally made a great effort, and as he took off each piece of clothing he noted down exactly where he had put everything that he had just taken off. The next morning the man arose, and pleased with himself, he took the slip of paper in hand and read: "cap"—and there it was, and he set it on his head; pants—there they lay, and he put them on, one leg at a time. So it went until finally he was fully dressed. The man turned to the mirror. "That's all very well, but now, where am I myself? Where in the world am I?" You see, all he knew was the clothes he put on, and not the man on whom the clothes were put.

I think today we are not so stupid, but in the schools we keep lists nonetheless behind which we sometimes duck. Lists that tell us what to wear, where to go, what to eat, when to sleep, what to play, what to remember, what to know, about what to laugh, and about what to cry. These lists tell us about what to think. But there are no lists in the schools that would help us know where I am myself, where in the world am I? And there are no lesson plans that would aid us in helping our children discover themselves. We ought to stand proudly and significantly in the classroom teaching our students where they presently stand, and on whom to put the clothes.

Study and Benevolence

In *Ta'anit*, a tractate of the Babylonian Talmud, the Rabbis speak to the condition of crisis—unnatural rains or drought, the presence of armies in the cities, fearsome animals roaming through the streets, buildings falling down, and plants growing unnaturally. For these events, the Rabbis said, they would "sound the alarm immediately," institute a public fast, and begin a process of repentance that might hopefully relieve these dangerous conditions. In the winter of 2002, when I spoke to the Curriculum Studies Project at Louisianan State University, I talked there about a crisis in the field of teacher education. I meant then to address the failure of institutions responsible for preparing teachers of creating a climate of rigorous intellectual activity that students could observe and absorb and aspire to model. I spoke then of an absence of inquiry and a dearth of intellectual challenge in academia. This talk became a chapter in *Talmud, Curriculum and the Practical: Joseph Schwab and the Rabbis.*[1] I assign this text regularly to graduate students.

In that winter of my discontent, I sounded the alarm immediately. I spoke of the crisis in the academy the seriousness of which ought to require the institution of a public fast. I said then that we have yet evidences of serious malaise and ineptitudes in our educational institutions, indicated not by low test scores but by low intellectual expectations; epitomized not by intellectual inadequacies, but by avoidance of intellectual challenge. Our schools were characterized not by intellectual rigor nor active pursuit of meaning and meaning making, but by intellectual retreat and cravenness. In the academy today, I accused, at the primary, secondary, and postsecondary levels, teachers have ceased learning for the ease of mere teaching. We have eschewed intellectual questioning for the safety of the answer. We wait anxiously for the word from governmental agencies for our deliverance.

In that winter of my discontent, I was steeped in my work on Joseph Schwab. In 1969, Schwab had decried the scarcity of models of intellectual communities then available to students to nurture their intellectual activity. His accusation thirty years ago still has relevance today. Schwab (1969, 18) complained that "the faculty have no professional lives apart from their teaching. They make no music. They write no books. They uncover no new knowledge. They forge no new policies. They are not conspicuously engaged in honorable public service. They administer little apart from their homes and classrooms. They teach, to be sure, but their teaching is a full-time service they perform, not a flowering or a sharing of expertise of scholarship..." Schwab accused "the visible bulk of the faculty [at larger universities] of being very busy indeed but not at their own business. They are not doing what flows from their talents but what is marketable..." (18). Students lacked a model of intellectual passion and honesty. Finally, Schwab bemoaned, in many places "there is an intellectual community, but the students don't belong to it, not even as second-class citizens" (18–19). Schools neither require nor invite students to participate in intellectual pursuit; rather, students are assigned minimum standards and evaluated on their achievement of that minimum standard. Schwab complained that the undergraduate curriculum "is a mere inculcation of...a rhetoric of conclusions and of a body of rote methods for solving rote problems" (19). Having spent almost half of my life in the schools, I concurred with Schwab's portrayal of the academy, and I decried the current dearth of intellectual endeavor and challenge. At that time, I sounded the alarm immediately. As has been my practice over the past several years, I looked to the Rabbis whose inquiries and conversations were edited into the Talmud. I looked to their questions for questions, I searched in their thinking for methods, and I sought in their wisdom for wisdom. In *Ta'anit* the Rabbis offered stories of teachers whose prayers could relieve the crisis, and I suggested then that were we to do our work, and not attend to the mean-spirited bureaucrats and politicians, we might address the perilous conditions besetting our beloved academy.

Sound the Alarm Immediately

It is another crisis I mean here to address, though this crisis manifests itself differently in the current educational scene. And I mean once again to sound the alarm immediately. I want to speak of the atmosphere of crisis under which teachers daily function, and of the stance teachers take in the face of the heavy critique from various pockets in the social fabric. We read regularly of the failure of the schools to educate our children; William Bennett (2001) has declared that "In America, the longer you stay in school, the dumber you get in relation to your peers in other industrialized nations."[2] We read regularly of the ineffectiveness of teachers and curriculum; we hear regularly the politicians' rhetoric bemoaning the horrid state of our educational institutions and the impending national doom that must inevitably follow from this failure. We hear regularly about the necessity of repairing these institutions immediately.[3] And in the educational communities, we have come to accept these accusations as true. We accept the criticisms of us and our work, and we deny the effectiveness we know we possess; we accept their descriptions of our incompetence, and agree to their retributive measures of correction; we accept their judgment concerning our failures and relinquish our authority to function as we know best. We give up our faith in ourselves and our work, suppress our love and awareness of the children, and in fear of the political bureaucracies, cower in the undercrowded teachers' rooms from the overcrowded classrooms. We subsist in a constant state of crisis. We cease to do our work.

The heart of the No Child Left Behind Law is not better education but standardized testing; the heart of the No Child Left Behind Law is not education, but accountability; the heart of the No Child Left Behind Law is not the welfare of the child but the reputation of the politician. As William Pinar (2004, 163) states, "Accountability is the American face of fascism." The ultimate intent of the No Child Left Behind Law is retributive—it is written to produce failure and then to punish what it has created. For example, in answer to the Frequently Asked Question (39), "Why do we need reading first?" the government responds, "Fewer than one in three fourth-graders are deemed 'proficient,' and by twelfth grade only forty percent of senior are proficient in reading for their grade."[4] Obviously, one must conclude, schools and teachers are at fault. This view interests me because on the Progress on International Reading Literacy Study, ten-year-olds in the United States finished ninth among thirty-five nations, and the scores of these top nine nations were so close that not many countries scored significantly higher than did the United States. The 1992 study *How in the World Do Students Read* shows the U.S. children finishing second. But the Bush administration decries the substandard reading performance of our children. So much for scientific research and statistical accuracy!!

Listen to the government's advertisement for the NCLB law: "It is built on four common-sense pillars: accountability for results, an emphasis on doing what works based on scientific research, expanded parental options and expanded local control and flexibility" (3). The aforementioned "accountability for results" refers to scores on standardized test. The "rigorous" scientific research itself is severely limited in scope—by definition that research already defines what it tests. As Dewey argued more than one hundred years ago, the vice of predetermined goals defeats the whole enterprise of education. And besides, there are contradictory studies suggesting alternative results, as in the two reading results listed above. And because students will be tested annually, parents will measure regularly their children's scores—assuming, we know, that parents understand those scores—and then evaluate the schools' performance based on that of their child. The law is built on an assumption of inadequacy: the present government claims that "The NCLB law is a landmark in education reform designed to improve student achievement and change the culture of America's schools." President Bush accused, "Too many of our neediest children are being left behind." And, I would note, forty-three million of our citizens—at least half of whom are children—have no access to health care, but I hear no blame cast upon the medical community for its failure to ensure the health of those children. On whom was Bush casting blame when he said that our children are being left behind? Indeed, all of the components of the law address perceived inadequacies within the school, and point to no failures outside of it. The present government claims that "No Child Left Behind does not label schools as failing, instead schools are identified as 'in need of improvement,' and they are given assistance to improve. Testing called for in NCLB helps schools identify subject areas and teaching methods that need improvement. For example, if reading scores do not reach the state's benchmark, the school knows it needs to improve its reading program." Ah, were it all so simple, and that schools existed in a social vacuum, and were not in some significant way products of virulent practices of segregation, gender bias, violence, misplaced political priorities, and savage inequalities! It will be a great day for education, the poster read, when schools have all the resources they need and the Air Force must hold a bake sale to fund the manufacture of a new bomber.

But despite my spleen, I do not here mean to martial a critique of that insidious law. I would, indeed, be better occupied. No, here I do not intend to address the law's stupidity. I would not spend my effort addressing that law, because that law distracts me from the life's work in the schools where, as Piaget reminds us, there are children all about us. I would not spend my effort addressing that law as if its tenets hold even the slightest tinge of truth or address a present reality. I would not expend my effort defending myself. I think often of Henry David Thoreau. In *Walden*, in the very first chapter, "Economy," Thoreau muses about the pedants who concern themselves with the facts and details that mark the educated person: "Many are concerned about the monuments of the West and the East—to know who built them." But, he adds, "For my part, I should like to know who in those days did not build them—who were above such trifling." I, too, would not spend my time studying these absurd laws; rather, I should like to be above such trifling, and I would like to place myself among those who in these days partook in neither the law's construction nor its implementation. And as for those who write and pass these laws as if they actually knew that about which they pontificate, and who raise up this law as their legacy, I look to Thoreau again for comfort. He writes, "As for the Pyramids, there is nothing to wonder at in them so much as the fact that so many men could be found degraded enough to spend their lives constructing a tomb for some ambitious booby, whom it would have been wiser and manlier to have drowned in the Nile, and then given his body to the dogs." It is this stance of defiance that I would take toward the law and its legislators—and urge all teachers to do the same. Its implementation degrades our lives, and it would have been manlier and wiser to throw it and its architects into the Mississippi.

No, I really do not wish to speak of the NCLB; there are voices far more eloquent than mine who decry what Gerald Bracey calls this "Trojan Horse." Of course, I cannot address an issue in public education today without addressing the NCLB law because there is no talk in education except of this law. And *this* is the crisis of which I wish here and now to speak. The crisis for which I would sound the alarm immediately stems from our willingness to believe the accusations of the politicians, to cower in the halls fearful of speaking, and to abandon our competences for the ease of following the established scripts of public officials whose scope of vision does not extend into the schools themselves. The crisis of which I desire to speak finds us again not being about our business, of betraying our selves and our purposes.

All Israel Shall Have a Share in the World to Come

Chapter 10 in *Sanhedrin*, a tractate of the Babylonian Talmud, begins: "All Israel shall have a share in the world to come...except he [*sic*] who maintains that resurrection is not a biblical doctrine, [that] the Torah was not divinely revealed, and [he or she who is] an epikoros" (90a).[5] Rabbi Akiba adds to this list of those who will not have a share in the world to come those who read uncanonical books—who give credence to false beliefs, or take time from proper study—and those who whisper a charm over a wound, thus purporting to imitate the God who heals. Chapter 10 interests me because unlike so much of Talmud, this chapter is comprised not of elaboration of the law, but of stories that pertain rather to the culture of Judaism. In most of Talmud, the Rabbis are concerned with elaborating the conduct of the activities of which daily life is comprised, of one's daily *actions* in this world. The justification for this focus is God's *Levitical* (19:2) command, "As I am holy, you shall be holy." Upon that standard the Rabbis organize daily existence to ensure the attainment of that holiness.

In much of Talmud, the focus is not so much on belief—that appears to be a given—but on action. But uncharacteristically, in Chapter 10 of *Sanhedrin* the Rabbis are, in fact, concerned with one's *beliefs* in this world; in this instance, the Rabbis argue from their belief in the resurrection, a life after death. To give particularity to these beliefs, the Rabbis offer the character of particular men—three Kings and four commoners—whose actions were deemed so heinous as to deny them access to the world to come. For the Rabbis, the actions of these men betrayed, and hence, endangered the existence of the entire culture. All of Israel is said to have a share in the *world to come* except those who are accused of not truly belonging to the culture *in this world*, whose actions betrayed and endangered that culture. These actions could have been motivated only from a failure of belief.

These particular heinous persons, three Kings—Manasseh[6], Jeroboam,[7] and Ahab[8]—and four commoners—Balaam,[9] Doeg,[10] Ahitophel,[11] and Gehazi[12]—are deemed guilty of undermining the culture of Judaism and threatening its integrity and its future. It is not important to the Rabbis if that individual is a person of rank—a King—or a person without privilege—a commoner. Everyone is equally obligated: those who are in positions of power and those who are daily workers must equally assume responsibility for their eternal future as a result of their actions today. The betrayal of public trust by each of the Kings betrayed the essential beliefs that defined the culture, and by their actions evinced a lack of faith in God. And each of these commoners attempted to betray the culture for his own personal ease and gain despite the awareness that by his acts he defied God's word and the central beliefs of the culture. That Doeg is said to be an Edomite and that Balaam is described as not of the people of Israel suggests also that the world to come is reserved for everyone and not only for the Jews. The exclusion from the world to come speaks to a person's actions and not necessarily to their avowed beliefs: Ahab, after all, remained a Jew throughout his life. Furthermore, the actions undertaken by these persons endangered the future of the entire Jewish culture, and they are thus, no less culpable of idolatry than those who deny the world to come; this world is no less significant than the next.

I am curious, of course, about the culture of teaching, and of those who would for their comfort and private gain betray that culture. Shall it not be equally said that to abandon the culture of teaching—the life of the mind's growth and the growth of the mind's body—is to abandon not only the culture of teaching but the culture in which teaching exists? These three kings and four commoners may each find their counterpart in the current crisis environment in the public primary and secondary schools and in the Academy of Higher Education. These three kings and four commoners may find their exemplars in teachers who abandon their purpose and betray the trust invested in them.

But, in truth, I am not here going to address this issue. The culture of teaching requires more elaboration than I have here time or inclination to offer. Indeed, it is my purpose here to give support to that culture and to those who function within it. I wish to suggest that the culture of teaching can only be strengthened, and thereby made more appealing, by an attention not to those who built the pyramids. I would speak now of more ennobling matters. I would like to address, then, what I believe to be the Rabbis' intent in their manifesto, an intent that is, I believe, directed at behavior—and to address the context of their dictum—the motives and effects of their directives, to see if there might not be in them some solace for teachers. If we are to maintain our beliefs, we must have something to do other than constructing tombs for boobies; we must have something else to do than that which does not ennoble our present lives or ensure our welcome into the next.

As I have said, in Chapter 10 of *Sanhedrin*, the Rabbis seem concerned not with what we must do, but what we must believe. And this Mishnah suggests that what must be believed is the certainty of the resurrection of the body. The Rabbis demand a singularity of belief, a prescription anathema to my liberal sensibility and the entire practice of democracy. I do not mean to excuse the authoritarianism of the Rabbis, and their seeming unwillingness to tolerate disparate thought. Actually, the insistence with which the Rabbis demand consistent belief bespeaks the presence of a significant body of dissenters to their opinion. The lady *doth* protest too much methinks. And finally, it is not so much belief but practice to which the Rabbis' dictum is addressed. The Rabbis seem to know that all belief has materialization in practice, and that all practice can define belief. I think it is not singularity of belief but of behavior that the Rabbis demand. And that behavior is meant to define Judaism as the Rabbis would define it.

During the years in which the conversations that were later themselves canonized into the Talmud took place, the Rabbis struggled to save Judaism if not from absolute extinction, then certainly, from absolute dissolution. The destruction of the Temple by the Romans in 70 CE and the impossibility of the Temple service meant that the entire structure of Judaism had to be reinvented if Judaism were to survive. We find in the Talmud the struggle in which the Rabbis engaged to transfer the rites and rituals of the Temple into the homes and the synagogues and schools. The Rabbis replaced the priests, and prayer and study substituted for the sacrifice. In the days following the destruction of the Temple, and in the loss of the ritual sacrifice as a central event, the Rabbis were fighting for their very existence and the continuing viability of Judaism. In their determination to define a right belief, they were also trying to define the behavior that would, in a world dramatically changed and seriously dangerous, develop from that belief.

I believe that Chapter 10 is consonant with the Rabbis' efforts, and that the discussion concerning resurrection and canonical books attempts to solidify the challenged authority of the Rabbis and to reestablish a culture of Judaism in the absence of the Temple and the structures and beliefs and rituals that were organized by its presence. The Rabbis were attempting to reinvent Judaism, and there was no dearth of critics and alternative Jewish sects vying for hegemony: the Karaites, Sadducees, and Essenes to name three. There was no lack of enemies who would advocate for the disappearance of Jews and Judaism, such as the Romans and the newly minted Christians. Neil Gillman (1999) asserts that the Rabbinic notion of resurrection, which has no authority in Torah, served to offer redemption to the martyrs of the Roman persecution in the first years of the new millennium—about the time of the destruction of the second Temple in 70 CE, and to threaten ultimate punishment for those soldiers who committed these atrocities. The Rabbis' insistence that the resurrection was reward *and* punishment was meant to give some strength to Jewish belief and continued Jewish practice in a continually hostile world. The availability of resurrection to all also undercut the challenge to Judaism from Christianity whose Jewish martyr was claimed to have risen as evidence of his divinity. Finally, the existence of resurrection would enhance God's power to absolute status because death's power is defeated by the idea of Resurrection.

Chapter 10 in *Sanhedrin* justifies belief under even the most horrid of circumstances, and reigns in that dissent that threatens to break the fragile developing culture. Chapter 10 promises ultimate reward in the next world not only for the soul, but also for the body that suffered horribly in this world as well. Chapter 10 in *Sanhedrin* offers relief to the horrors in this world by suggesting that though unknowable here, God's ways are not unjust. Chapter 10 in *Sanhedrin* ensures that continued ethical practice in this world finally has its reward, though not necessarily in this world, and that heinous behavior has its punishments, though not necessarily in this world. The maintenance of this belief addressed in this chapter would serve to maintain the ethics of daily practice: I am holy therefore you should be holy. If, for the most part, Talmud is the explication of what that holiness entails in the exigencies and minutiae of daily life, this chapter is the foundation of that practice in the assertion of God's justice. It is an early version I think, of the owl eyes of chaos theory—in time, order will be understood, though none of us are ever afforded that much time. Nonetheless, we must act *as if* the world depended on our behavior.

Now, belief in the world to come makes possible belief in the Messiah, the national redeemer. It is the Messiah whose arrival will save the nation Israel ("On that day I will raise up the fallen booth of David..." [*Amos,* 9:11]), and it is the Messiah who will usher in the era of individual resurrection. I am not here interested in attesting to the actuality of the coming of the Messiah. Indeed, the actual nature of the Messianic age intrigues me; though its reality is vague, it is a consummation devoutly to be wished. Its arrival is crucial to our times. But I would like to address briefly several comments by the Rabbis' concerning the conditions that may (or may not) attend to the time of the Messiah's arrival. I would like to address the Rabbi's arguments concerning the world's condition at the moment of the Messiah's arrival and see if perhaps the Rabbis have something to offer us in education.

If the coming of the Messiah promises a surcease of suffering, then this arrival must be certainly sought. How might the Messiah's advent be recognized? The first mention in Chapter 10 does not bode well for us in teaching, though it may portend the imminent arrival:

> In the generation when the son of David [i.e., Messiah] will come, scholars will be few in number, and as for the rest, their eyes will fail through sorrow and grief. Multitudes of trouble and evil decrees will be promulgated anew, each new evil coming with haste before the other had ended.   (97a)

How prescient the Rabbis' description now seems to me. We do not live in happy times: the rains have stopped and there are unnatural plant growths in our fields; wild animals and soldiers roam in our cities and our buildings fall down. According to the Rabbis, the Messiah ought to arrive soon. The alarm must be sounded. We are in much need of redemption. But alas, I see no signs of the Messiah's arrival.

For the Rabbis, when the Messiah will arrive remains very much an issue. In their discussion in *Sanhedrin* (97b), the Rabbis tell stories of the Messiah's expected arrival. And they wonder under what conditions the Messiah will arrive—it is, of course, a pressing question. They would know if the time is at hand, or at least, how they might recognize the time when, indeed, it will be at hand. Elijah, the Prophet, a personage steeped in legend and association with the coming of the Messiah, says to Rab Judah the brother of R. Salla the Pious, "The world shall exist not less than eighty-five jubilees, and in the last jubilee the son of David will come." *Not less,* says Elijah, but perhaps more. And when Judah asks Elijah if the Messiah will come at the beginning of the final jubilee or at its end, Elijah says that he does not know. There is hope expressed, but little anticipation. And Rab Ashi adds that the Messiah is not to be expected before the eighty-five Jubilees have come, but after they are passed, the Messiah may yet still come. "Afterwards, thou mayest await him." Waiting is the game here. And what shall we do while we wait?

Waiting

There is another related story in *Bava Metzia*.

> Elijah would frequently appear at Rabbi's academy. One day it was the first day of the month, [and] he was delayed and did not come. He said to him: "What is the reason that you, sir, were delayed?" He said to him: "[I had to wait] until I awakened Abraham and washed his hands, and he prayed, and I laid him down, and similarly with Isaac, and similarly with Jacob." "But you should have awakened them together!" "I thought [that] they would pray fervently, and would bring the Messiah before his time."

Elijah, the very same Elijah who left sufficiently vague the date of the Messiah's arrival, is a familiar figure in Jewish culture; historically, Elijah is the Prophet from Gilead. (I think of Poe's poem, *Lenore*: "Is there no balm in Gilead?" The legends of Elijah suggest that there is, indeed, such succor.) Now, Elijah was the outstanding religious leader of his time. The Bible records that he did not die, but was carried to heaven in a chariot pulled by horses of fire. Among the many stories told about the Prophet Elijah, perhaps the most important ordains him as the forerunner of the Messiah. In this tradition, Elijah is charged with devising the coming time aright. This Talmudic story quoted above rests clearly in that Messianic tradition. I will return to this matter shortly. But it intrigues me that this Elijah is portrayed first, as a common servant and second, as a regular visitor to the Rabbi's studio. Could it be that even Elijah continues to study? And is his study, perhaps, connected to the coming of the Messiah? The centrality of study in Judaism, its placement at the center of faith, argues for this position.

Another aspect of this story intrigues me: Elijah explains that he has taken the more inefficient route in his actions because the conjoint praying of Abraham, Isaac, and Jacob would have been so fervent that they would have brought the Messiah before his time. Now, the Messianic theme speaks of the coming of redemption and the advent of unlimited peace and happiness. The Messianic era portends the end of exile, of suffering, and of political and economic strife. The Messiah offers the redemption for which the Jew waits. Not unlike Beckett's Vladimir and Estragon, who wait for Godot, the Jew waits in great expectation and hope for the Messiah. Yet, oddly enough, in this story, but unlike the desires of Vladimir and Estragon, Elijah is fearful of bringing the Messiah before his time—he would rather, as it were, keep Godot off stage and thereby, continue to await his arrival. In this Talmudic text, waiting is a consummation devoutly to be wished.

It is often written that Elijah can assume many appearances to accomplish his purposes, but here, however, Elijah appears as a common citizen familiar to Rabbi Yehudah Ha-Nasi and as a regular visitor to his academy. It intrigues me that Elijah would be so regular a visitor here—again I wonder, could it be that the Talmud suggests that study is intrinsic even in the lives of the prophets—and that even the great Prophet Elijah must continue his study at Rabbi's academy. Elijah is, we are told, a daily visitor.

Second, it is interesting that in this story Elijah is portrayed as servant to Abraham, Isaac, and Jacob, the traditional founding male ancestors of Judaism. He awakens them each and every day and cares for their diurnal needs. The patriarchs pray each day—it interests me that Elijah does not appear to pray with them; rather, he is wholly preoccupied with the exigencies of their daily lives. Could it be that the Talmud suggests that the great Prophet Elijah who will usher in the messianic era must live a quotidian life until he is called upon to announce the coming of the Messiah. As there is a story that the Messiah exists even now among us, it is clear that Elijah, too, must exist now among us. It is he or she who ministers to our daily lives. Who makes it possible for us to pray, or in more contemporary terms, as it were, to study? Elijah comes to our studios—to our offices—daily. I wonder what might it be like if we treated every visitor to our offices as if they were the Prophet Elijah? As if they had just come from ministering to Abraham, Isaac, or Jacob—or Sarah, Rebecca, Rachel, or Leah—performing necessary duties executed separately for each to ensure that their conjoint praying not bring the Messiah before his time? Indeed, what if we considered ourselves to be Elijah so mindful of our responsibility to others that we put off our own prayers and studies until we had completed our ministrations to the Other? I suspect that were we to think of each other and ourselves as this Elijah, the time for the coming of the Messiah would, indeed, be here.

And why not awaken the patriarchs at once? Certainly it would prove to be the more efficient method. And it might be argued, wouldn't now be an appropriate time for the Messiah's coming? But Elijah says that the conjoined prayers of the patriarchs would bring the Messiah before he or she is due! The coming of the Messiah for whom we wait is here contingent on human action. That is, in this story the Messiah will not usher in the era of peace but will arrive at its moment. According to Talmud, the Messiah, paradoxically it would seem, must wait for us. Of course, we might say that

Godot waits for Vladimir and Estragon, but they do not see it this way. And the play itself twice disappoints Vladimir and Estragon when the boy announces that Godot will not come today. "Let's go," Estragon says. (*They do not move.*)

There are two arresting possibilities concerning this story of Elijah. The first is that the coming of the Messiah will not bring redemption but rather, acknowledge it. That is, human action must prepare the world for the Messiah. The second possibility is equally as astonishing: that the Messianic era is only possible by human action. Were the patriarchs to pray in unison the Messiah might arrive, but the world would be as yet unprepared for the Messiah's coming. There is, it would seem, very much something to be done.

Study and Benevolence

When will the Messiah come? That the Messiah will come seems never in doubt, but in the meantime, we must yet continue to act. What should be the work done while waiting? In *Sanhedrin* (98b) Rabbi Alexandri points to an apparent contradiction in Torah raised by R. Joshua b. Levi. Torah, the sacred text, may contain no contradiction, and therefore, Alexandri resolves what appears an apparent contradiction in the text. Rabbi Joshua b. Levi had noted that it is written in *Isaiah* 60:22, "I, the Lord, will speed it in due time." Alternatively, these lines may be quoted, "I am God, in its time I will hasten it." The Rabbis assert that these lines refer to the coming of the Messiah, but Joshua b. Levi had pointed out that if the Messiah will come in the Messiah's proper time, when the Messiah must come, then how is it possible that God will hasten that arrival? And Rabbi Alexandri answers that the lines mean that *if* the people are worthy, *then* God will hasten the Messiah's arrival, but *if* they are not so worthy, *then* the Messiah will come in his time. In either case, the Messiah's arrival is predicated on a world deserving of redemption—and since we cannot know when we are worthy, then we cannot cease working for that moment. This perspective seems consistent with the earlier story in *Bava Metzia*, and argues for a world in which our work aspires to make us worthy of redemption.

But the question persists: what should that work be while we await the Messiah's arrival? In their discussion in *Sanhedrin,* the Rabbis turn to Scriptures to identify the time of the Messiah's arrival. Rab asserts that "The son of David will not come until the [Roman] power enfolds Israel for nine months." To arrive at this conclusion, Rab interprets a line from *Micah* (V:2): "God will deliver [Israel] to its enemies until the time that a woman in childbirth gives birth; then the rest of his brothers will return with the children of Israel." Rab interprets that the Messiah will come only after the Roman power has subjugated Israel completely— that is, in all of those lands to which the people of Israel are exiled— for as long as a woman is pregnant—nine months. Ulla responds, and Rabba agrees, "Let [the Messiah] come but let me not see him." Abaye wonders aloud, what might be Rabba's reason for not wishing to see the coming of the Messiah. Could it be, he wonders, that Rabba does not wish to see the birth pangs—the difficulties that precede the birth of a new era? And one of Rabbi Eleazar's disciples asks, "What must a man do to be spared the pangs [which precede the coming] of the Messiah?" And Eleazar answers, "Let him engage in study and benevolence."

It is a curious answer. But I think it addresses the crisis in education we daily confront. We must wait for no deliverer, though there are no end to the claims of false Messiahs: Ralph Tyler, Madelyn Hunter, E.D. Hirsch, William Bennet, George W. Bush. Things are breaking up out there, Reb Dylan tells us. I think that he is right; our buildings are falling down and crumbling in decay. We are besieged by soldiers. We teachers must wait for no deliverer. We teachers must ourselves be our comfort. For it seems to me that the Rabbis have taught us what we must do while waiting; a waiting that itself would create conditions for the Messiah. We teachers must be ourselves our saviors. We must study and perform acts of benevolence. I think that this study is not a quietistic retreat into the ivory tower of the academy; rather, it is a study that is linked to acts of benevolence. When we ask what the Rabbis mean by benevolence, we discover that it is not indistinct from study. Resh Lakish says that "he who teaches Torah to his neighbour's son is regarded by Scripture as though he had fashioned him." As his proof text Lakish (99b) goes to Genesis. There it is written that when Abraham left Ur he "took his wife Sarai and Lot, his brother's son, and all their wealth that they had amassed, and the souls which they had made in Harani." Lakish argues that because Abraham taught these pagans about the singular God, it was *as if* he had created them himself. In teaching is creation.

R. Eleazar offers an alternative interpretation of the importance of teaching: he claims that to teach Torah to his neighbor's son is regarded in Scripture as though the teacher herself had created the words of the Torah. As his proof text Eleazar quotes *Deuteronomy* (29:8): "Keep therefore the words of this covenant, and make them." To teach, says Eleazar, is to create what is taught. And Raba adds a third interpretation. He says that when one teaches Torah to the neighbor's son it is "As though he had made himself, for it is written, 'and make them': render not *them* but *yourselves*." Raba argues that the Hebrew word for "them" can be variously rendered as "yourselves,"[13] and therefore, the verse can mean that to teach is to create the self. Finally, R. Abbahu says that he who causes his neighbor to fulfill a precept—he who teaches how to fulfill the commandments—is regarded by Scripture as though he had done it himself. For his proof text, Abbahu offers *Exodus* 17:5. There it is written that "the Lord said unto Moses...take...they rod, wherewith thou smotest the river." But Abbahu notes, "Did Moses then smite it? Aaron smote it! But he who causes his neighbor to fulfill a precept is regarded by Scripture as though he had done it himself." In this interpretation, there is nothing but the doing of benevolence and it derives all from studying and teaching.

What should we do in the midst of the birth pangs of a new world? And the Rabbis' answer is simple—engage in study and benevolence. Thought and action. Tikkun olam. We must be about our business.

Part 3

Every Third Thought Is My Grave

Chapter 7

Intimations of Immortality:
An Ode of Sorts
(With apologies to William Wordsworth)

I am finishing my sixth decade. Every third thought shall be my grave.

On the eve of his fatal battle with Macduff, Macbeth ponders his portion:

> My way of life
> Is fall'n into the sere, the yellow leaf;
> And that which should accompany old age,
> As honour, love, obedience, troops of friends,
> I must not look to have; but in their stead,
> Curses not loud but deep, mouth-honour, breath
> Which the poor heart would fain deny, and dare not.   (V.iv.22–28)

There is at that moment not all the balm in Gilead that would comfort Macbeth. His life's character has withered, dried up, and shriveled; what once promised to bear fruit has become spiritless and dead. It could be Macbeth if it were not himself about whom Dylan (1997) writes, "Every nerve in my body is so vacant and numb / I can't even remember what it was I came here to get away from." All the feeling Macbeth can summon at the news of the suicide of his once-beloved wife is "She would have died hereafter." Macbeth has no children, and the legacy he leaves behind despises and condemns him. Life, he says, is a tale told by an idiot, signifying nothing. In this, the end of my sixth decade, I ache for Macbeth's losses. I worry about his conclusions.

Alas, I have killed no king, but I now wonder of what that honor, love, obedience, and troops of friends that might accompany me might consist as I end this decade. As dear Madeleine might ask, what traces am I leaving? I have two wonderful children whom I cannot burden with my legacy, own (as long as it is mine) a mortgage, and have written thus far five books with a small (even a *very* small) academic readership. I appear in remarkably few bibliographies, and I have stood with several thousand students and gazed into their sometimes bored and sometimes defiant faces. I am now pondering what my life might signify. I have been in the classroom for thirty-five years, and I am wondering if there are curses not loud but deep, which speak of me, and if I am the subject of mouth-honours only.

Rabbi Joshua ben Levi said, "In twenty-four places we find that the *beth din* [Rabbinical Supreme Court] inflicted excommunication for an insult to a teacher, and they are all recorded in the Mishnah" (*Berakhot*, 19a). I am certain the Rabbis thought more of teachers than does the present in which I live. I should have been a contender. For thirty-five years in the classroom, I have learned and taught, I have taught and learned. And what is it that I have done? In 1973, when I was in graduate school at the State University of New York at Stony Brook, the students and faculty had a memorial service for Pablo Neruda. On the cover of the stenciled program was a fragment of a poem only the last line of which I remember. It read: "I am not part of this crime."[1] There are times I wish to absolve myself of responsibility for the state of the world, for the state of my students, and for the neuroses of my children, but Hillel calls me back: "If I am only for myself, what am I?" I muddle on. As I finish my sixth decade, I am reviewing the situation.

When Hamlet stabs wildly through the arras thinking to kill the king, his mother cries, "Hamlet, what have you done?" and he answers, "Mother, I know not." Sometimes, I know what it is *I do*, but I cannot always answer for what it is *I have done*. But, in time, perchance, I might nose it as I climb the stairs. I recall the terror in Billy Pilgrim's mother's voice (Vonnetgut, 1969, 38) as she lay on her deathbed and despairingly wondered, "How did I get so old?" The simple answer to her query was that she had gotten so old because she had lived so long, but in fact, the question stemmed from a greater existential situation: what happened to me? I am wondering that now. If Eliot Eisner promises me an immortality, I would know now of what that immortality consists.

I am finishing this year, my sixth decade. I am, as it were, reviewing the situation, and wondering what happened to me. Every third thought shall be my grave. I have been long wondering what Prospero, in Shakespeare's *The Tempest*, meant when he announced his plans for his return to Milan. I think that it is also to his retirement in Milan to which Prospero retires himself. Many of my friends have retired from their classrooms after long years of activity in them. Sometimes I feel that I alone am left to tell the tale.

On this last night that Prospero will spend on the island, he invites Alonso, the Duke of Naples,

> To my poor cell, where you shall take your rest
> For this one night; which, part of it, I'll waste
> With such discourse as, I not doubt, shall make it
> Go quick away. The story of my life,
> And the particular accidents gone by
> Since I came to this isle; and in the morn
> I'll bring you to your ship, and so to Naples,
> Where I have hoped to see the nuptial
> Of these our dear-belov'd solemnized;
> And thence retire me to my Milan, where
> Every third thought shall be my grave.   (V.I.300–311)

Prospero's offer to tell Alonso the story of his life, and in this telling to make the night pass quickly, suggests a life of great excitement and interest, but by his own admission, the reality seems to be that Prospero lost his kingdom to his brother, Antonio, exactly because he was so immersed in solitary study that he did very little to engage in government and life, and, de facto, gave up his control. "I thus neglecting worldly ends, all dedicated / To closeness, and the bettering of my mind / With that which, but by being so retir'd, / O'er-priz'd all popular rate" (I.ii.89–92). Prospero by devoting his life to withdrawal and study, "to my state grew stranger, being transported / And rapt in secret studies." I wonder what Prospero could now tell Alonso that would keep *him* up all the night in rapt attention. Prospero's own daughter, Miranda, was not at all interested in her father's tale; when he attempted to narrate to her his story, she lost focus and attention. She fell asleep! Ah, but perhaps she had heard the tale before. Alas, I would think that Prospero's would be a very solitary, sedentary, even plodding story, and one not at all inclined to help pass the long night. And yet, Alonso seems eager to hear the tale. He states, "I long / To hear the story of your life, which must / Take the ear strangely." I think that it is not the story per se, but how Prospero's learning adds to that story that is crucial here. It must be a new story he will narrate, and not the one he attempted to tell Miranda. It is a different Prospero who will tell this tale, and it interests me that with all of his study, it is not to the moral, but to his story that Prospero returns at the end, not to the didactive but to the narrative. It is not, it would seem, what he *knows*, but that *he* knows that makes all the difference.

I have told stories all of my life, and, I fear, put to sleep more than a few Mirandas and Ferdinands. I am finishing my sixth decade, and I would waste some part of this night to tell such discourse as might make the night pass quick away. Perhaps it is not the *story* I tell, but that *I* tell the story that might help us make it through the night. But, finally, how do we tell the dancer from the dance? I am, of course, my stories. Perhaps I, along with Prospero, have had to learn that. We are all storytellers, and we teachers make our lives from it. I narrate a teacher's story; the teacher narrates his or her life. I would invite you to my poor cell, where you shall take your rest for this one night; which, part of it, I'll waste with such discourse as, I no doubt, shall make it go quick away.

It is very quiet out here in the cabin. There is still an hour left to darkness, and only the light of the desk lamp and the computer screen sheds any illuminance. I am writing out here, telling the story I create from the life I live outside here. In here, I create the life I live out there. It is peaceful in here. Often, I feel like Huckleberry Finn: "We said there warn't no home like the raft, after all. Other places seem so cramped up and smothery, but a raft don't. You feel mighty free and comfortable on a raft." Ah, there is no home like the cabin, I feel so free and comfortable out here. I have three chairs: one for solitude, two for company, and three for society. I do not permit the cats to enter Walden.

But, alas, the raft floats on the river, and the river has no preferences, and moves inexorably on pushing Huck and Jim willy-nilly. It brings tragedy, separation, and the King and Duke. If they are to succeed in their quest for freedom, Huck and Jim must learn to tell a different story so that they might live their lives off of the river. Rather than give Jim back to the Widow Watson, Huck *on the river* redefines himself by choosing actions he might not have made on the land: "I took [the letter] up, and held it in my hand. I was trembling, because I'd got to decide, forever, betwixt two things, and I knowed it. I studied a minute, sort of holding my breath, and then says to myself: 'All right, then, I'll go to hell'—and tore it up." Huck writes a new story by tearing up that old one, and neither of them are the stories he planned. But the new story is necessary if he and Jim are ever to leave the river. If Huck had never left the raft, how would he ever have learned what understanding he ever did acquire about the world? I know the world is too much with us late and soon, though out here in the cabin, the world and the cats seem to remain outside the closed door. Eventually, I must open it, however, and all but the cats enter and my isolation is disturbed. I will leave the cabin for a while, and I will rewrite the story I have composed in it.

We must each and all tell the story. But what story shall we tell if, like Prospero, in study we neglect "worldly ends, all dedicated / To closeness, and the bettering of my mind / With that which, but by being so retir'd, / O'er-priz'd all popular rate." As for the pursuit of worldly ends, the world has long prevented the acquisition of any significant accumulation in that quest, and kept the worldly ends of teachers quite minimal. As for those *other* worldly ends, actual purposes in the world, this would be scann'd. We teachers prepare people for the world, but I am wondering what we actually know of the world for which this preparation is mandated. I wonder how we teachers study that world, and understand our presence in it. What stories do we teachers tell ourselves? First, perhaps, we must learn something for ourselves that might make our nights go quick away.

Thoreau says that we would always rather be living than writing about our lives, but I think that if we never write, then we never narrate our lives, and perhaps we never quite have a life to live. In the case of the latter, we exist only in the moment, and in the absence of time. There, we are only reactive; in the unreflective present we are not our own but time's agents. We act unthinkingly from a past that is always a present. We act as we have always acted; we need not consider. We are not story, because there is no narrative, and without story, we lack all agency. Outside of the stories we tell, we are not cognizant of our influence and power. Certainly, we can in this world beat our children or our spouses, we can act on our rage and hate; in the world, our actions have consequences. But if we cannot narrate the story in which consequences exist, then we are imprisoned in the present. Stories, after all, have structure, and that structure contains an end. Our stories have the power not to stave off death, but to make it absent from our minds. Our stories would make the night pass quick away. What is the relationship between the scholarship I practice out here in my Walden, the scholarship I espouse at the university, and the stories I tell in the classroom? If my learning is always first for me, if when I study it is me who is fulfilled, then how can I stand face to face with students? And if I do not study for me, what can I offer them? If I do not study, then who am I? If I do not study, I have no story to complicate. If I do not study, then who will? What have I offered them over the years? I am not certain I know how to answer any of these questions.

I am a storyteller myself. For the past thirty-five years, I have been a teacher in the public schools of New York and the public university of Wisconsin. I have practiced curriculum and I have taught curriculum. I am a curricular theorist. I am a storyteller myself. I've been wondering of late what I might say about those stories I've told. I suspect that like Prospero, my instrument is predominantly Reason, though I suspect, too, that it is not exercised through the grace of God—but more of that later. Nor can I use my Reason, as did Prospero, to liberate the spirits of nature and control its brute forces, though perhaps, like Spinoza, I might use it to deal with my own passions. Pinar (2004, 247) writes that as curricular theorists "our tasks as the new century begins is nothing less than the intellectual formation of a public sphere in education, a resuscitation of the progressive project, in which we understand that self-realization and democratization are inextricably intertwined. That is, in addition to providing competent individuals for the workplace and for higher education, we must renew our commitment to the democratization of American society, a sociopolitical and economic process that requires the psycho-social and intellectual education of the self-reflexive individual." I think this is a project that is implicit in Prospero's story. "Come," says Tennyson's Ulysses, " 'tis not too late to seek a newer world." I tell a story.

But for now as I finish my sixth decade, I am curious to know what Prospero means when he states that upon his return to Milan "every third thought shall be my grave." Actually, for years I have mistaken Prospero's words; for years I believed that he said "every third thought shall be *of* my grave." I could better understand Prospero if my misapprehension of his statement was not so flawed, and that Prospero had, indeed, said that in his retirement every third thought should be *of* his grave.[2] That would have made sense. As Prospero approached death, his grave would be more in view than his birth; the closer to the grave the more prevalent in his thought would it be than thoughts of the long-ago cradle. As his activity decreases, so might increase the thoughts of his physical decline. As he approaches the grave with studied awareness, his thoughts might reflect that knowledge.

But, alas, that is not what Prospero says: he states clearly that every third thought shall *be my grave*. It could not mean that each thought is a dead thought, nor even that every thought is of death. A grave is the repository for a dead body, and he says that every third thought will be his grave. Not the thought's grave; it is not dead; rather, it will possess him for eternity. Prospero would take every third thought as his grave—his final resting place, thoughts, perhaps, from which he may not retreat, and are no longer available to revision. (But then, I consider, two out of three thoughts are *not* his grave, and there is more to learn.) Perhaps Prospero indicates that there is a material component to every idea; an idea is of the earth and holds the thinker as a grave would hold the body. I think also that Prospero suggests that every third thought leads him toward the grave and thus becomes his grave, for in this imperfect world of mortal beings, our thoughts do end, oftentimes sooner than we would like. Perhaps Prospero speaks of thoughts that no longer change, and hence, are dead. These would be regrets, though I do not sense that Prospero holds these. I devote no energies myself to regrets; with Spinoza, I seek only joy. Mistakes and errors lead not necessarily to regrets; rather, Prospero acknowledges them as inevitable and valuable. Finally, maybe Prospero refers here to final thoughts into which he will wrap himself for eternity. But first, Prospero would tell Alonso a story that would consume the night.

I suspect that I am Prospero's age; Shakespeare was fifty-two years old when he died. I hear times' winged chariot. I would not think *of* the grave, though alas, I do, and I am pondering now what it means that every third thought *is* my grave. What are my thoughts? I study out here in my cabin-office every day, and I have life yet to live. What does thinking about Prospero tell me about what I am doing? Every third thought is of Prospero's every third thought, and I am not unconcerned with those other two-thirds. If I can define those thoughts and that grave, perhaps I, too, can narrate my life.

And where is immortality here?

Rabbi Uri (d. 1826) taught: "If you stand in front of a tree and watch it incessantly to see how it grows and to see how much it has grown, you will see nothing at all. But tend to it at all times, prune the runners, and keep the vermin from it, and in good time it will come into its growth. It is the same with man: all that is necessary is for him to overcome his obstacles, and he will thrive and grow. But it is not right to examine him every hour to see how much has been added to his growth."

Perhaps I have not watched the tree as it has grown, though I think I have pruned the runners and kept the vermin from it. I have overcome a few obstacles, and have achieved my growth. I'm reviewing the situation and examining my growth.

Intimations of Immortality: An Ode of Sorts

Each morning, I arise early (only that day dawns to which we are awake!), and in the dark and quiet of this new day, I stumble down the stairs to the kitchen. The house is very quiet, though the old floors and beams creak and groan in harmony with my stiffening frame. The cats rouse up sleepily from their sleep-places, look at me puzzled and with not a little amazement and, I imagine, even a little disdain, and then return to their slumber. I move slowly down the stairs grasping the banister with one hand, and with the other, push against the wall. I feel somewhat balanced and secure. I follow the beacons of night lights as I would those emergency path lights on airplanes that I am assured will lead me to safety in the event of a loss of power and a crash into a dark place, and head toward the kitchen and the promise of a taste of redemption.

I grind the beans, boil the water, and carefully prepare my morning coffee.[3] This ritual comforts and sustains me. Its practice acknowledges the presence of the day,[4] affirms my sense of self, and confirms my expectation for the future. My morning coffee reminds me of yesterday when I had stood before the stove and the whistling kettle, and this coffee leads me to tomorrow when again I will measure the shiny beans into the burr grinder; my morning coffee settles me in the present, as I pour the boiling water through the fine dark, brown powder, and the aroma of the brewing liquid fills the room. This morning is, of course, somewhat changed from yesterday, but ironically then, in this respect it is the same as yesterday, which, too, was different than the one before. But this cup of coffee, in its difference, is like that of no other day. It is *today's* beginning. I open the cabinet containing the coffee mugs, I make a play at choosing, though inevitably, I select the same mug daily, and I splash a dollop of half-and-half into the dark, pungent, and steaming liquid. I'm ready. I gather my eyeglasses, my books, and papers, put them in the Reader's Tool Box, and head out into the near dark to my cabin-office. Pushing the door open, I turn on the light and set my box and mug down. I take a cautious sip; it tastes of immortality.

In the modestly lit spaces of my simple cabin-office settled behind my house, I read the morning headlines in my online version of the *New York Times* seated in front of the computer screen. This reading is another ritual I practice. Though I have lived outside of New York City for many years, the *New York Times* is how I learned to read newspapers. With all of its flaws and warts, it is still the daily I want to read, though these days, as I have above noted, I look at only the headlines. I want to be assured that the next tragedy has not yet occurred. When I am certain, or, almost certain, that all remains today calm, rather, was yesterday calm, I turn to the obituaries. Who would have thought death had undone so many? At this moment, I inevitably think of Weaver member, Lee Hays, who wrote the following: "I get up each morning, and dust off my wits / Open the paper and read the obits / If I'm not there, I know I'm not dead / Then I eat a good breakfast and go back to bed." I search for familiar names, and graph my life with those whose bar had traveled with mine but now has suddenly ceased. I examine carefully the ages of those who have died in the past twenty-four hours, and I perform some quick (for me, who is mathematically challenged) arithmetical operations comparing the age of the deceased to my own present condition. I am taken aback at how close I am in lived years to too many of the deceased, and I am not comforted by my seeming longevity. Oh, it is a morbid business, and if I didn't know better, I'd take some medications and go back to bed. But once again, I'm not anywhere on these pages. I take another sip of coffee (it is a fine cup I have brewed today!), and I am ready to do something. I open the current document and write the first word. I am immersed in ritual. I address the world. I am immortal. I love rituals; they comfort and sustain me. I think that it is these rituals that present me to the world, and, I think, make me present in it. Participation in rituals offers me, too, I think, a taste of immortality. Partaking in rituals, I share the experience of generations that precede me, and I prepare for those yet to come: the children. Rituals that I have established (like the preparation of the morning coffee), or rituals that I have accepted (like attending *shul* on Shabbat, or entering the classroom daily) ground me in the day, in my communities, and in time. Rituals are also my situation in the immortal. For me, these rituals are not the well-worn paths that led down to Walden and from which Thoreau finally turned, nor are these rituals patterns of conditioned response accomplished without thought, like tying my shoes. Rather, rituals are regular activities that in the accomplishing set borders and boundaries for the day and for the seemingly boundless and incomprehensible universe. My rituals offer me a frame in which I might create the day; they define the canvas on which I may produce my painting (see Milner, 1950). Rituals are requisite for my ability to do anything in this world; they are the grammar that gives meanings to my symbols.

Participation in rituals engages me in immortality because rituals engage me sustainedly in my world. Rituals are obedience to laws that we practice in this world, and therefore, are those events that lead us to that world. It is commonly held that to live forever finally free of the world is the hope for immortality. Spinoza (1955, 269) remarked that "Most people seem to believe that they are free, in so far as they may obey their lusts, and that they cede their rights, in so far as they are bound to live according to the commandments of the divine law. They therefore believe that piety, religion, and, generally, all things attributable to firmness of mind are burdens, which, after death, they hope to lay aside, and to receive the reward for their bondage, that is, for their piety and religion." However, it is just these burdens, these activities associated with piety and religion, *these rituals,* by which one *achieves* immortality. *Conscious* participation in the world, and by conscious I mean participation with an understanding of the connections between our ideas and their bases in the world, imperfect as that understanding may be, immerses us in the eternal. "To know something, to have a true and adequate conception of it, is to understand how it came about and why it is as it is and not otherwise" (Nadler, 2006, 59). When we act according to this understanding, we are the freest to act, because we act according to the way we are learning things are, rather than by someone else's perspective of the way we simply wish them to be.[5]

Rituals are our activities based in our adequate ideas. For Spinoza, only God possesses adequate knowledge of everything—all ideas are in God, he would say—but the more our ideas approach adequacy, then the more we reach beyond our finite condition and become *like* God, who exists, of course, outside of time, in eternality. Since for Spinoza (1955), the mind is the idea of the body,[6] then "He who possesses a body capable of the greatest number of activities, possesses a mind whereof the greatest part is eternal" (267). The greater our activities in this world, the greater are our ideas; and the greater are our adequate ideas, the more we approach God's knowledge that is immortal.

Immortality is, thus, achieved in the present and in practice. Spinoza says, "And, since human power in controlling the emotions consists solely in the understanding, it follows that no one rejoices in blessedness, because he has controlled his lusts, but contrariwise, his power of controlling his lusts arises from this blessedness itself" (270). We are most free not when we deny ourselves; rather, our freedom to deny our lusts derives from our activities. Freedom is responsibility, and responsibility is immortality. "Existence of this kind is conceived as an eternal truth, like the essence of a thing, and therefore, cannot be explained by means of continuance or time, though continuance may be conceived without a beginning or end" (46). The free man thinks least of all of his death: immortality is not to live forever in time, but to live forever without time.

Not the private compulsive acts of modern neuroses (though I suppose they might, indeed, sometimes be that), I hold, rather, that rituals are active responses to the demands of the world that is always filled with the Other. Michael Morgan says that "Ritual acts are part of the mythology of theology; they deal with Jews and God, but what they accomplish is an important step in the acknowledgment of the face of the other person" (Morgan, 2005, 10). Obeying the ritual law of an unseen God requires us to see beyond the invisible to that which can be seen. Morgan says (10) that for Levinas obedience to the ritual law "expresses or trains the Jew to see what lies hidden in concrete, everyday life, a dimension of responsibility, purpose, and meaning often neglected or obscured or wholly occluded." Ritual affords us a vision of the Other. For Levinas, ritual in Judaism is "a response to divine command," but in that response, the Jew acknowledges and responds to a beyond or Other, ostensibly God, but as Levinas sees it, "the face of the other person" (10). In the performance of rituals, God's presence becomes known; knowing God is understanding the word and fulfilling the ritual. In his essay "To Love the Torah More Than God," Levinas says (in Kolitz, 1995, 30), "The guarantee that there is a living God in our midst is precisely a word of God that is not incarnate. Trust in a God who does not reveal Himself through any worldly authority can rest only on inner clarity and on the quality of a teaching." That teaching is the ritual law: this you must do. We look to the absent God, and we see only Others; we follow the ritual law, but in the absence of a corporeal God, the ritual prepares us to respond to and to serve the Other who is always present and visible before us. God's law teaches us how to care for the Other. Participation in ritual acknowledges a beyond and an Other. The response that ritual demands is our responsibility. Immortality consists in my activity in the world serving the other. Participation in ritual assures my immortality.

To follow ritual, then, is to acknowledge that we are aware of a summons by the Other, and therefore, that we recognize the vulnerability of the Other. We are ourselves Other and, therefore, vulnerable. "Thus," Morgan says, "Jewish ritual trains the Jew in being responsible and not just in being aware of [responsibility] in some cognitive sense." Ritual is not about belief, but about practice. To stand before the other in ritual is to accept responsibility for that other. Morgan writes, "Every social encounter—no matter what its content or character—is *always already* a response to the other person and an act of acknowledging that person" (11). In fulfilling ritual, we obey the command of the unseen other. Not formalized behaviors, but responsibilities, rituals are responses to the unspoken command of the other. As rituals immerse me in the immortal, immortality is forever my action in the present. Insofar as we understand this, we understand ourselves as Spinoza would say, *sub specie aeternitatis*, under the aspect of eternity: it could be no other way. From this understanding derives our pleasure and our immortality. As Roger Scruton (1999, 28) explains, "By gathering our chains into ourselves, and becoming conscious of their binding force, we also rid ourselves of [the chains], and obtain the only freedom that we can or should desire." When we accept our responsibilities to the Other, we achieve our freedom. It is the function of our learning that we achieve our freedom. The teacher's immortality exists in his or her engagement to ritual.

Ah, but the teacher's days are so now regularized, so set from without. These are, perhaps, no longer rituals, responses to the Other, rather, they are the exercise of pro forma practice. They inevitably fail at immortality. In Talmud (*Berakhot*, 28b), the Rabbis address aspects of prayer.[7] Jews engage in prayer three times every day, and the Rabbis address the issue whether ritualized prayer negates its function. The Rabbis have already set the order of, and the script for prayer, but they are concerned with the formalized aspect of it. "Rabbi Eliezer says: "If a man makes his prayers a fixed task, it is not a genuine supplication." Too routinized, Rabbi Eliezer suggests, prayer loses meaning; if our behaviors are accomplished by rote and without thought, then they lack content—they are not ritual. Rabbi Jacob ben Idi adds in the name of Rabbi Oshaiah that "Anyone whose prayer is like a heavy burden on him" does not utter a genuine supplication. The obligation of experience without willing intent oppresses. Here, too, prayer is not ritual. Today we would describe such a person as depressed, and prone to take to his or her bed; that individual would not rise into the world to stand before the Other. In this case, activity could be imprisonment.

But two other Rabbis, Rabbah and Rabbi Joseph, offer a different perspective on ritualized prayer, and one that opens it to responsibility and activity. They say that "If one cannot add something fresh to the prayer, then they do not utter a genuine supplication." That is, for Rabbah and Joseph, the ritual provides the frame for behavior, and within those borders the individual is free to improvise.

Too much of our classrooms are routinized and lack ritual, and I suspect that the Rabbis would not accept the effort there as prayer. Prayer and study are expressions of awe; prayer and study are ritual, and sacralize the mundane. When we pray and when we study, we take a stance in awe and humility, and we actively acknowledge, as Abraham Joshua Heschel (1959, 52) says that "our lives take place under horizons that range beyond the span of an individual life, or even the life of a generation, a nation, or an era." Prayer and study emanate from the silence of awe and wonder; they are rituals upon which our present acquires presence.

Walter Schuller, an Amsterdam physician, wrote that "It seems that death's unexpected debilitation took [Spinoza] by surprise, since he passed away from us without a testament indicating his last will" (in Nadler, 1999, 340). Well, of course, he did and he did not. Spinoza had written in the *Ethics* that "a free man thinks least of all of death, and his wisdom is a meditation on life, not on death" (1955, 232). In the *Ethics*, Spinoza acknowledges that "A free man is one who lives under the guidance of reason, who is not led by fear, but who directly desires that which is good, in other words, to act, to live, and to preserve his being on the basis of seeking his own true advantage, wherefore, such an one thinks of nothing less that of death" (232). It is not that death does not exist for the free man; rather, that man is so wholly engaged in life that he never even considers death. But Spinoza knew that complete freedom did not come easy, nor was it available to all, or even to any one, really. Even he, I suspect, suffered from some lack of freedom.[8] Spinoza did, after all, live in this world, and like the rest of us, did so with imperfect understanding. That is, he did not always possess that understanding of the essence of things and of the infinite number of causes that determine our volition. "The things which I have been able to know by this kind of knowledge are as yet very few," he admits (1955, 9) in the *On the Improvement of the Understanding*, because most of the things we know come through experience and not the intellect, and therefore, our ideas of external bodies, of our own bodies and minds are mostly inadequate. We are not wholly free when we do not live wholly on the basis of reason. Spinoza says that to live in obedience to virtue—which is nothing other than to act, to live, and to preserve one's being—requires obedience to reason. If a man does anything contrary to reason, and as a result of inadequate ideas, that man is passive because he does something that does not follow from his virtue. "But," Spinoza says, "in so far as he is determined for an action because he understands, he is active" (204). As our rituals engage us in willed activity, these rituals must follow from our virtue: "Since that which we, in accordance with reason, deem good or bad, necessarily is good or bad; it follows that men, in so far as they live in obedience to reason, necessarily do only such things as are necessarily good for human nature, and consequently for each individual man; in other words, such things as are in harmony with each man's nature. Therefore, men, in so far as they live in obedience to reason, necessarily live always in harmony with another" (209). What we would want for ourselves, we would so desire for others.

But, the human being who is not so free does not always comprehend that order nor live in it, and therefore acts on inadequate ideas, and lives in human bondage. Spinoza says, if a man "be thrown among individuals whose nature is in harmony with his own," then certainly this man (and these men) should live free and naturally, *unless*, of course, that person were to be thrown with those not at all in harmony with his nature, in which case he will hardly be able to live without making accommodations." This latter case, Spinoza acknowledges, is the lot of most men, including, I would think, Baruch Spinoza. We must all make our compromises, thankfully so. Nevertheless, Spinoza adds, "[W]e shall bear with an equal mind all that happens to us in contravention to the claims of our own advantage, so long as we are conscious, that we have done our duty, and that the power which we possess is not sufficient to enable us to protect ourselves completely; remembering that we are a part of universal nature, and that we follow her order" (1955, 242–243). If we acknowledge that we are not alone in the world, but have tried to live the life we have dreamed, we can not only acquit ourselves, but understand our place. It was actually the Rabbis who first said something not unlike this statement from the heretic Spinoza, whom the Rabbis excommunicated in 1656. Rabbi Tarfon would say, "It is not your responsibility to finish the task, yet you are not free to withdraw from it" (*Pirke Avot,* 3:21). It strikes me that the task set here is that of understanding ourselves and living in harmony with responsibility for others. The frame that we might set about this goal is ritual.

We have in education a pedagogy that aspires toward this freedom: *currere*. Madeleine Grumet (1976) describes *currere* as "a reflexive cycle in which thought bends back upon itself and thus recovers its volition" (130–131). *Currere* is the continuously evolving story cycle of our lives.

Currere, the verb and not the noun that derives from it, offers a pedagogy that places us essentially in the story in which we place ourselves. Thoreau says that no matter where we are in the world, when we look up at the stars, we are at the center of the universe. *Currere* teaches us, from our centering, about the world we narrate with ourselves as its center. *Currere* teaches us how to complicate the story we tell, and offers strategies for understanding the complexities of the stories of others.

I think this is exactly what Spinoza spoke about in *The Ethics*. We are human, he teaches, and the power we possess is not sufficient to enable us to protect ourselves completely from the complexities and exigencies of living with other people. We are, alas, human and therefore, he says, "liable to passions…that every individual wishes the rest to live after his own mind, and to approve what he approves, and reject what he rejects" (Spinoza, 2005, 270). As we better understand our passions, and teach others to better understand theirs and those of others, we gain and teach freedom. This understanding is the intellectual work of Reason, and might be achieved in the performance of our daily educational rituals. Pinar (1976) writes, "As we work politically, socially, and educationally, we are working psychically…we are educated to the extent that we are conscious of our experience and to the degree that we are freed by this knowledge to act through skills required to transform our world" (21, 38). Spinoza argued this hundreds of years earlier. In a letter to G.H. Schaller, Spinoza writes that a false freedom is a "human freedom which all boast that they possess, and which consists solely in the fact, that men are conscious of their own desire, but are ignorant of the causes whereby that desire has been determined" (1955, 390). Spinoza asserts that it is the work of reason to lead us to reason. "To act absolutely in obedience to virtue is nothing else but to act according to the laws of one's own nature. But we only act, in so far as we understand (III.iii): "therefore, to act in obedience to virtue is in us nothing else but to act, to live, or to preserve one's being in obedience to reason, and that on the basis of seeking what is useful for us" (1955, 204). The Rabbis argued in *Eruvin* (10b) that "If the Torah had not been given we could have learnt modesty from the cat, honesty from the ant, chastity from the dove, and good manners from the cock who first coaxes and then mates." For the Rabbis, Torah is their engagement in study and reason, and from it they could achieve our freedom and our immortality. We have other texts.

Currere offers a method of education and freedom. Rituals, of which schooling and curriculum are two, provide the frame; reason the method. Spinoza writes,

> For the ignorant man is not only distracted in various ways by external causes without ever gaining the true acquiescence of his spirit, but moreover

lives, as it were unwitting of himself, and of God, and of things, and as soon as he ceases to suffer, ceases also to be.

Whereas the wise man, in so far as he is regarded as such, is scarcely at all disturbed in spirit, but, being conscious of himself, and of God, and of things, by a certain eternal necessity, never ceases to be, but always possesses a true acquiescence of his spirit." (1955, 270)

Immortality is engagement in the world. Ritual is the means of engagement. Performing rituals, we achieve immortality. The free man is not free of the world; rather, he understands it. He can tell stories about it. Education should inspire in us the desire to be virtuous and to live an ethical life.

The free man need make the least accommodations. I am not so free. Every third thought is of the grave.

And what is this ritual I daily practice? The Jew obeys ritual law that demands attendance to the Other. Teachers, of which I am one, attend to the Other daily. Ritual is how it is that I come to stand in the classroom each and every day. I open the door. I command them to command me. I work until I deserve them. We teachers might think of our work as ritually enacted. The teaching demands we care for the stranger in our midst, and ritual law teaches us how to do so. The ritual *prepares* us to do so. Levinas (1990, 18) says that "The way that leads to God therefore leads *ipso facto*— and not in addition—to man; and the way that leads to man draws us back to ritual and self-education. Its greatness lies in its daily regularity." It is when we attend to our students that we are engaged in ritual. Leon Brunschvicg says, "One can only work effectively for the future if one wishes to realize it immediately" (44). I think this is accomplished in ritual. Rabbi Judah Ha-Nasi agrees with Ben Pazi, who asserts that the verse that contains *all of Torah* reads, "You will sacrifice a lamb in the morning and another at dusk." Here, Rabbi acknowledges that engagement in ritual requires daily effort, and that daily effort—behavior and not belief—focuses always on service to the Other. Why should God care about our lambs in the morning and at dusk except that this sacrifice of ourselves sustains the world and creates the world to come. Our daily rituals taste of immortality.

It is this faith in the ritual law that Zvi Kolitz (1995, 22) affirms in "Yossel Rakover Speaks to God." From the midst of the embattled Warsaw Ghetto, Yossel Rakover says, "I love Him, but I love his Torah more ... God means religion, but His Torah means a way of life, and the more we die for this way of life, the more sacred and immortal it becomes." That way of life is obedience to ritual law: sacrificing a lamb in the morning and another at dusk. Perhaps, for Judaism, ritual law offers sufficient reason for that trust, though the presence of ritual does not guarantee one's participation in it. That ritual law—perhaps *all* ritual is law—directs us outward and toward the Other. Ritual leads us out into the world where the Other must be faced. Levinas (1990) writes that "The ritual law of Judaism constitutes the austere discipline that strives to achieve this justice" (18). Ritual law "is effort. The daily fidelity to the ritual gesture demands a courage that is calmer, nobler, and greater than that of the warrior" (19). We engage in ritual as the practice of our daily lives, and we engage in rituals *as if* we are summoned. For Levinas, what always summons us is the face of the Other. That face commands. "For Judaism, the world becomes intelligible before a human face," writes Levinas (1990, 23). Rituals are a response to a divine command; rituals are undertaken in "deference to God," but since God is incorporeal and has no face, then the ritual leads us to the face of the Other. Engagement in ritual

> . . . is either a step on the way to, or an expression of the relationship of the self being summoned and commanded by the other person; it facilitates or expresses one's recognition of the other's epiphany or call. Such ritual acts, which help to train Jews to acknowledge otherness and hence to recognize responsibility to others, do so by occurring in nature and by setting up "distance" between the Jew and nature, on the one hand, and God, on the other. (Morgan, 2005, 10)

Thus, engagement in ritual becomes an ethical stance: it is responsibility. "Ethics is an optic, such that everything I know of God and everything I can hear of His word and reasonably say to Him must find an ethical expression . . . The attributes of God are given not in the indicative, but in the imperative," says Levinas (1990, 17). To know God is to know what must be done. When we engage in ritual, we stand in ethics, and we know not only God but ourselves as well. Self consciousness, self-education, is an awareness not of ourselves alone, but of ourselves before the Other. Self-consciousness is a consciousness of justice and injustice. To know the harm that the Other experiences is to know my consciousness. As Spinoza might agree, this is to stand in immortality.

Education, Levinas (1990) teaches me, is "obedience to the will of the Other" (18). In this interesting sense, all education is self-education: I learn what it is that *I* must do; *I* learn how to listen to the even unspoken demands of the Other, and then *I* obey. I teach as I learn. But, sometimes, my classrooms have been so still, and I have stood in despair and anger at the silence: I felt lost and alone. But I think now that perhaps I was not listening. Too often, perhaps, I stood in the classroom and I misapprehended convention for ritual. When I thought things were going well, perhaps it was myself of whom I was thinking; perhaps the most ritual classes were the ones in which I felt most obliged and the least satisfied. Our current pedagogies are oriented—even mandated—toward declamation and not perception; our curricula are designed for delivery and reception. There is little ritual attached to our practice.

And so, of my morning stance in the classroom, I wonder now what I have offered of myself that the students have not yet made request. If obedience to ritual law is to give way to the commands of the Other, and engagement in ritual is all self-education, then the classroom is where I have learned to obey and to act and to become immortal. The classroom ritual is what I follow to learn. I learn that I might follow classroom ritual, and I consider now that when I have taught in the public schools, when I have taught *well* in the public schools, it was never the content that improved my students, but how through the content they have learned to hear the command of the other. It was always ethics I taught, and it was when I stood in ethics that I was the teacher and immortal. Over the years, it seems now that my discomforts in the classroom arose from the conflict between following the demands of ritual that led me to the Other, and the demands of the curriculum that have been organized by final tests. Those practices I cannot call rituals do not lead me to the face of the Other, and are not responses to any command. Here, curriculum at its best is an imposition on the Other, and at its worst it is oppression. Classroom rituals might have improved my days.

To hear the command of the Other is not to avoid the rigor of the intellectual activity the classroom enacts. Rather, that command almost always says "Teach *me*! Help *me* learn!" And that command contains always the personal—help *me*, teach *me*. It is the teacher's stance to respond with the materials that will enable each, from his or her centering, to learn about the world that each narrates with ourselves as the center. *Currere* teaches us how to complicate the story we tell, and offers strategies for understanding the complexities of the stories of others. The teacher's rituals enable her to hear the command of every student: teachers are often physically and emotionally spent at these efforts. The immortality of the teacher resides in that exhaustion: engaged in our rituals, we prepare for the world here and for the world tomorrow. Our classroom rituals educate. They offer us immortality.

A friend came to visit Thoreau as he lay dying. "Can you see the other shore?" the friend asked. And Thoreau answered, "One world at a time." A free man, I recall, thinks least of all of his death, though of his death he sometimes thinks. As Dylan (1997) says, "I've been to Sugar Town, I shook the sugar down / Now I'm just trying to get to Heaven before they close the door." Entering my seventh decade, I am, as I have said, thinking at this time about this one world, and wondering about the next. Every third thought is my grave, and I am intrigued by what Prospero might have meant by that statement. Perhaps it is I am not so free as Spinoza, and perhaps Thoreau offers me another perspective on Prospero, on my life, on my teaching, and on immortality. Remaining doubtful of the immortality Eliot Eisner offers the teacher, I seek a hopeful alternative.

In his autumnal essay "Autumnal Hints," Thoreau addresses the closing season of a life; he had for sometime been ill with tuberculosis, and I suspect he felt his impending death.[9] The free man thinks least of all of his death, though of his death he sometimes thinks. In the essay, speaking of the leaves of Autumn, Thoreau (2001) writes, "They that soared so loftily, how contentedly they return to dust again, and are laid low, resigned to lie and decay at the foot of the tree, and afford nourishment to new generations of their kind, as well as to flutter on high! They teach us how to die. One wonders if the time will ever come when men, with their boasted faith in immortality, will lie down as gracefully and as ripe, with such an Indian summer serenity will shed their bodies, as they do their hair and nails" (382). Unlike Dylan Thomas' urgent call to "Rage, rage against the dying of the light," Thoreau offers a quiet acceptance of the naturalness of dying even as he advocates glory in the naturalness of living. If there is immortality in this death, it is not easy to discern in Thoreau's elegy.

Eisner (2006, 44) described the immortality of the teacher as an intermingling of lives, so to speak: "Their lives live in ours, and our lives live in theirs." I am not quite certain I find much relief in that. As I expressed earlier, there is a parasitism implicit in this residing, and I do not know if this powerless resting offers much hope in immortality; we have no knowledge of or agency in this residing; we are immortal almost by chance. I recall Spinoza's student's angry calumny, and I shrink from this immortality. I have spoken above of immortality residing in my daily activities, and I suspect that this is what Thoreau meant when he spoke of one world at a time.

There is, for Thoreau, apparently no distinction between the dying of the leaves and their living, their decaying at the foot of the tree and their fluttering on high: both, Thoreau asserts, are aspects of existence. Both processes are a growth. In Thoreau's paean to the real and metaphorical autumn, there is no tone of regret; it is as if in death that Thoreau discovers life. There is creative activity *and* beauty in living *as well as in* dying. As the leaves have fluttered high during life and their growth, so they now lay low in death and decay, and offer nourishment to the generations of their kind. The leaves are food for the next generation's growth, but the leaves do not continue to live in that generation. Remarkably, as food they would, as it were, disappear. If there is immortality, Thoreau suggests, it resides in the calmness and efficacy with which we can accept our deaths as essential to life. Thoreau's image of the Indian summer, usually considered as a period of sunny warm weather occurring in the autumn after the first frost and just before winter, as an unexpected blooming and a renewed relevance, speaks to a fecundity in the process of dying. Our decline is an ascendancy. Death, for Thoreau, though an end is not a finish.

And perhaps, here rests an aspect of immortality onto which I might take hold. Our calm deaths leave the world to the living, and our decay sustains them. My death is my life's end, and my life has been least of all about death. Immortality is the coming to life from a dull and spiritless death-in-life. "Everyone has heard the story which has gone the rounds of New England, of a strong and beautiful bug which came out of the dry leaf of an old table of apple-tree wood, which had stood in a farmer's kitchen for sixty years, first in Connecticut, and afterward in Massachusetts,—from an egg deposited in the living tree many years earlier still... Who does not feel his faith in a resurrection and immortality strengthened by hearing of this" (Thoreau, 2001, 447). Immortality consists in living, and depends not on our deaths. Even in their deaths, the leaves live. Looking out at Walden during the fall, Thoreau watches the brilliant display of autumn color in the leaves, and he notes that only the funeral bell had meaning, and that the other sounds were all so trivial and frivolous. "In proportion as death is more earnest than life, it is better than life" (1962, Vol X, 76). Our lives must contain that earnestness that the brilliant leaves of Autumn display if our lives are to have meaning. We teachers attain our immortalities in the brilliance of our lives rather than the leaving of it.

Thoreau suggests a different relationship between the old and the young, the dying and the living. In the quoted passage, he calls the activity of death one that affords nourishment, and one could argue that this nourishment refers to some spiritual inheritance. I do not think this is Thoreau's intent, however. One world at a time. The *Oxford English Dictionary* notes that one of the earliest meanings of the word "nourishment," from 1413 CE refers to "that which nourishes or sustains." In this sense, Thoreau notes that the decay of the leaves provides nourishment for leaves in generations to come. In the shedding of our bodies we provide that which nourishes or sustains, but clearly, it is our lives and not what we teach that sustains. Suggesting that erecting a lightning rod is an act of faith no less than deciding not to erect one, Thoreau says, "It only suggests that impunity in respect to all forms of death or disease, whether sickness or casualty, is only to be attained by moral integrity" (1962, Vol IV, 157). As Spinoza had said, our living ethically is freedom from death. Immortality derives not only from *what* I give, but from *that* I give. Dying is an activity of living, and my immortality resides in my real death and physical decay. Indeed, the earliest appearance of the verb "nourish" dates from 1290 CE, when it held the meaning, "to bring up, nurture, or rear (a person)." Teachers, of course, nourish; and in the Thoravian sense, teachers are immortal because they nurture, bring up, and rear their students. In giving ourselves up to students, we attain our immortality. It is not what is left, but what is used up in which our immortality exists.

Also in 1413 CE, "to nourish" meant "to promote or foster (a feeling, habit, condition, state of things, etc.) in or among persons." Here, to nourish was not at all the act of giving, but rather, the creation of conditions in which growth could occur. Here, nourishment is a sustenance, but not necessarily the material support. It is not *what* I offered as a teacher that sustains, though this is not incidental. Rather, nourishment consists in the fact that, as a teacher, I maintained a classroom in which growth could occur. Again, it is to this nourishing stance that Thoreau attributes our immortality. In this sense, teachers achieve immortality not in what we leave behind, but in what we are now. It is not to eternity that Thoreau sends these leaves or our souls; rather, it is to immortality and to this world that he directs all. In teaching us how to die, they teach us how to live, and it is in this living that we achieve our immortality. Finally, the Rabbis and Thoreau (1980) have always given me leave to play with word meanings[10] and I mean to follow this example. In 1380 CE, Wyclife employed the word "to nourish" to mean "to cherish," a meaning that, though now obsolete, holds great significance for me here. To cherish, to hold dear, to take affectionate care of a thing, to treat with affection and tenderness is what teachers might do in the classroom. In this sense, "to nourish" accords with the idea of ritual and responsibility; and transforms the idea of nourishment into the act of cherishing. In the ritual act, we attain our immortality.

Furthermore, Thoreau's (2001) redefinitions of our commonplaces hold meaning to my explorations of my immortality. The great harvest occurs, he says, not when I gather in what I have sown, but when I provide in my dying and my leaving for the living. Employing the language of the economist, Thoreau says that in their dying the leaves pay back the Earth for what they have received from previous generations of leaves. Of the fallen leaves of October, he writes, "This, more than any mere grain or seed, is the great harvest of the year. The trees are now repaying the earth with interest what they have taken from it. They are discounting. They are about to add a leaf's thickness to the depth of the soil. This is the beautiful way in which Nature gets her muck, while I chaffer with this man and that, who talks to me about sulphur and the cost of carting. We are all the richer for their decay" (381). The harvest here is not what is taken from the fields, but what is given to them by the death of the leaves. Death is the harvest of the year; not what *will grow*, or even what *has grown*, and may be now consumed or sold and traded, but what will no longer grow. Rather, this harvest of which Thoreau speaks is the yield of having lived. I think Thoreau offers a different perspective on immortality here. Eisner (2006, 44) has said that a teacher's immortality exists as images in the minds of students. He writes, "Those teachers past sit on our shoulders, ready to identity infractions and offer praise for work well done. Their lives live in ours, and our lives live in theirs." I am reminded here of Newton's statement that he could not have seen so far had he not stood on the shoulders of giants. I have always had narrow shoulders, and even in my well-heeled shoes, I stand only five foot eight and one-half inches tall. Eisner's statement also calls to mind Freud's notion of the superego, and I shudder to consider myself in that role. Neither of these images seem to approach Thoreau's meaning or his idea of immortality. For Thoreau, the dead do not live on in the living, which smacks to me of a form of parasitism. Rather, in dying, the dead nourish the living by being gone.

And whereas economists use the term discounting to mean the removal of interest from the principal *before* the loan is made, Thoreau asserts that the decaying, dying leaves pay back the loan *with* interest. They become earth. Discounting here seems an addition and not a diminishment. Immortality is what the dead give to the living by their deaths. In our deaths, we continue our engagement in ritual; in dying, we continue our responsibility, and we become immortal. Ah, to some we stink of sulphurous fumes of decay, and for others we become too expensive. Nevertheless, we are the Earth, and without us there could be no future growth.

Perhaps Thoreau's disquisition on autumnal hints is like Prospero's every third thought—it is clearly leaves about which Thoreau is concerned, but it is certainly death about which he speaks. But how Thoreau seems to adore the transformation of the living into the dying? "What is the late greenness of the English Elm, like a cucumber out of season, which does not know when to have done, compared with the early and golden maturity of the American tree? The street is the scene of a great harvest-home. It would be worth the while to set out these trees, if only for their autumnal value" (2001, 377). We would put our elders out for display, Thoreau argues, and not hide them away in assisted living quarters where they may die unseen. Thoreau was as free a man as I have met, and this essay points his way to dusty death. Here, however, there is no sound and fury, no tale told by an idiot signifying nothing. Rather, here is Thoreau's paean to death and decline. In death is our immortality; not in our having died, but in our having lived. The autumn glories are beautiful, and as beauty is in the eyes of its beholder, so those eyes must be taught to look for that beauty with the eye. The essay, "Autumnal Hints" teaches us to see in decay beauty, and to know that the beauty of the fall resides in the death of the leaves.

Here is a lesson that sustains the classroom. Perhaps, teachers might partake here of immortality by teaching immortality.

I think that Wordsworth's "Intimations Ode" speaks of this immortality. Though I do not hold here with Wordsworth that the child is father to the man (else what is an education for?), nor that on the child rest those truths "which we are toiling all our lives to find" (1967, ll.15–16) (I long ago ceased the quest for absolute truths waiting to be found, though ironically, I may be writing here of my faith in one), I hear in the poet's words an acceptance of our physical deaths, and the acknowledgment that in this "primal sympathy," what I name and accept as responsibility, and in our knowledge of the Other's need for us to relieve the suffering, we continue to live. We are immortal. It is in our deeds and ideas that we continue to live.

> What though her radiance which was once so bright
> Be now for ever taken from my sight,
>   Though nothing can bring back the hour
> Of splendour in the grass, of glory in the flower;
>   We will grieve not, rather find
>   Strength in what remains behind;
>   In the primal sympathy
>   Which having been must ever be;
>   In the soothing thoughts that spring
>   Out of human suffering;
>   In the faith that looks through death,
> In years that bring the philosophic mind.   (ll.175–186)

For Wordsworth, as for Thoreau and Spinoza, it is our lives that make us immortal and not our deaths.

> Thanks to the human heart by which we live
> Thanks to its tenderness, its joys, and fears,
> To me the meanest flower that blows can give
> Thoughts that do often lie too deep for tears.

Not here regret, but eternal beauty.

In the dullest of rooms, we could find beauty. In the dullest classrooms, we can teach beauty. In the dullest of classrooms, we can be beautiful as we beautifully decay. In the dullest classroom, we can become immortal.

*   *   *

There is this lovely story in *Berakhot* (18b) about which I have thought for years, and out of which now I might make some meaning. It concerns the relationship between the living and the dead, a subject about which I have expressed some concern. Every third thought is my grave, but every two thoughts remain here, and touch on the grade and the brave.[11]

Rabbi Hiyya's sons went out to cultivate their property, and in the midst of working, they began to forget their learning. Alas, such is too often the case with us teachers. In the process of getting a living, we forget our lives. We loved knowledge. We loved even the children, but some of us have allowed ourselves to become distracted and waylaid. Too many of us, perhaps, have become afraid, and refused to act in the best interests of the children and ourselves. Too often, we gave up our purposes to the managers and supervisors who usurped our authority. We have paid too much attention to the managers and supervisors and foremen. We have eaten well, and become, perhaps, too comfortable. We forget our learning. The story offers little clue whether the sons were just too preoccupied to hold onto their learning, or whether they carelessly let it go. To the Rabbis, I suppose, the cause is not as important as the result. In either case, the sons felt remorse over their loss, and they tried honestly to return to study, and to recall what they had once learned. During one of their learning sessions, one son asked the other whether their father, now deceased, knew of their situation. The brother answered confidently, "How should our father know? Remember," the son said, "in Torah (*Job*, 14:21) it is written, 'His sons come to honor and he knows it not; when they suffer he will not discern them.'" As the deceased cannot know of honor, then certainly they cannot know dishonor: thus, their father, now deceased, had been spared the pain of knowing his sons had forgotten their learning! But the other son, also quoting Torah (to which I suspect all three brothers have returned), responded, "How can our father not know what we have done, for doesn't Torah say, 'But his flesh grieves for him, and his soul mourns for him'" (14:22). Clearly, he concludes, the flesh of the dead can grieve and the soul can mourn; certainly the dead remain sentient and concerned with the living.

I think that it is the consequences for the dead and not for the living that the sons worry, for they are already in the process of recovering their learning. They are concerned that their father could know of the spiritual abasement of the sons. Perhaps he would interpret their loss, temporary though it may be, as his life's failure. His immortality here, so to speak, remains unquestioned, though the legacy of his life is unsure. Honoring the dead, here, is to continue in life, and, as Eisner says, his immortality resides in his residence in his son's minds.

I wonder, however, of what the father's failure might consist that calls his legacy into question. The brothers seem to seriously regret their loss of learning: in the fields, perhaps, they remembered study's richness and warmth. I recall Daniel Isaacson's comfort (Doctorow, 1971) sitting in the library and writing his dissertation. Amidst the stacks of books, he asserts, "I can go on." I feel immortal myself out here in the cabin with my coffee, my books, and computer screen. There is comfort in our learning, even though there sometimes seems so little result from it. There is comfort in our learning, and even though we believe that we possess what sustains, it is never enough. I study more. Every third thought is my grave, but perhaps the other thoughts bear some fruit. The Rabbis have asserted that learning that continues throughout three generations will never cease to exist. Perhaps it is for this loss that the father would grieve his sons' loss of learning: the end of learning. Perhaps for the Rabbis who tell this story, learning is eternal, and even, perhaps, eternity.

The Rabbis of the Talmud taught that when God created the world, God looked within the Torah for the blueprint. Within that text, God found the world. The study of Torah, then, is the study of our universe and of human existence. Of course, many of us no longer ascribe to this singular view of this singular text, but we are scholars, nonetheless, holding that the world *may be discovered* in our books. In their labors, perhaps, the brothers have lost the world as well as themselves. Perhaps every third thought is their father's grave.

In *Sanhendrin* (Chapter 10) the rabbis assert that all Israel shall have a share in the world to come, though what exactly survives into that world is not clear. If their father's soul exists and remains sentient, and if it yet possesses knowledge and concern for the living, then the brothers' loss of learning has consequence for the father, now deceased. He would grieve and suffer for their loss of learning, though again, it is not certain whence that grief stems. And if the dead retain sentience and can be affected by the living, then what respite is there in death when we have shuffled off that mortal coil? Hamlet worries about the character of the next world, but if all Israel has a share in the world to come, then the Rabbis promise an undiscovered country that is not frightening. But, if all Israel has a share in the world to come, then it is unclear to me what would trouble the brothers here. Again, it would seem that it is themselves of whom they think, and as Eisner says, their father's immortality consists of their conscience rather than their legacy.

But I recall, in Torah the dead remain dead, in their graves, so to speak, and there is no talk of a life after death. We know where the bones rest of Abraham, Isaac and Jacob, Sarah, Rebekah, and Leah; we know where Jacob buried Rachel. Everyone who dies in Torah (except the Prophet Elijah, and perhaps Serah)[12] remains dead and buried. Moshe buries his sister, Miriam and his brother Aaron, and Aaron buries his two sons. Their bones remain in the grave. God buries Moshe but it is clear that neither he nor his like will ever return again. Of this the Torah is insistent: "So Moses, servant of God died there, in the land of Moab, by the mouth of God. God buried him in the depression, in the land of Moab, opposite Beth-peor, and no one knows his burial place to his day." *But Moses was buried.* And the text continues, "Never again has there arisen in Israel a prophet like Moses" (*Deuteronomy*, 34:5–6, 10). Moses' life ends with his death. Thoreau, too, has suggested in "Autumnal Hints" that the dead stay dead, and that their immortality resides in their physical decay. They nourish the living by being gone; it is in their decomposition that they nourish. As they are no more, so they live. Eisner (2006, 44) has said that the teacher's immortality rests in the images of us that is retained by students, asserting that "Those teachers past sit on our shoulders, ready to identify infractions and offer praise for work well done." But that situation does not seem to be the source of the brothers' guilt; they have already identified their own infractions, already begun to rectify their condition, and only wonder if their dead father knows of their lapse. I wonder what difference his knowledge could possibly make to them. Perhaps it is their own pain to which they refer. Perhaps it is their own lives of which they speak.

Our students leave us regularly, and very, very few ever return. Of those who have forgotten their learning, we know nothing. Of what affect does this have on me, I wonder. Mostly, I think, none. If I work for the future, then that future is, indeed, unknown. And my present is wholly unaffected by it.

Interestingly, no more is heard from the brothers, but the sentience of the dead to the living is of some interest to the Rabbis. They seem concerned that immortality might bring its own sets of miseries. I have of late been curious about this Rabbinic interest, as every third thought is my grave, and I wonder what that means. I have dealt in "Study and Benevolence" with the Rabbis' concern with the afterlife; here I am concerned with why the Rabbis might be concerned not with the fact of the immortality of the soul that they assume, but with the relationship between the dead and the living. It is my concern as well. I am finishing my sixth decade; every third thought is my grave, but I have two other thoughts as well.

This particular story of the brothers, and the one that follows concern to what extent the dead feel the pain *of* the living. To the Rabbis, the relationship between the dead and the living must be important. The Rabbis' concern suggests that our obligations are not only to those before whom we stand face to face today, but also to those before whom we must stand eternally. For Spinoza, the wise stand always in eternity that exists only in the present. He does not seem at all concerned with obligations to the dead; they are indeed, out of mind. And Thoreau gives little attention to the dead. One world at a time. At best, the dead are compost. But these Talmudic stories suggest that our deaths do not seem to end our concern for the living, though the nature of that concern seems to be the question of the stories. This relationship between the dead and the living is prefigured in Deuteronomy 29:13–14, when Moses says to the people of Israel,[13] "Not with you alone do I seal this covenant and this imprecation, but with whoever is here, standing with us today before Adonai, our God, and with whoever is not here with us today." The Rabbis have always interpreted the final verse to refer to the binding nature of the covenant between God and the people for all the generations that followed after these desert wanderers. One need not have stood at Sinai to be part of the covenant.

Perhaps this is the basis of the Rabbis' concern here in their stories. Those that are now dead have fulfilled their responsibilities to those now living, and the brothers are concerned that they are no longer fulfilling the covenant. All of our behaviors—of the living and the dead—are in fulfillment of the covenant—you shall be to me a nation of priests and a holy people: to care for the widow, the orphan, and the stranger in our midst—for the Other. Our responsibilities to the dead are to pursue this goal for the living. In teaching, we fulfill our obligations not so much in *what* we teach—you shall be to me a nation of priests—as in *that* we teach. In telling these stories, the Rabbis suggest that we act not in individual or temporal isolation; rather, we must act here in this world *as if* our deeds affect eternity.[14] Both Thoreau and Spinoza would concur here.

The Rabbis argue about the source of the sentience of the dead. Rabbi Isaac says that to the dead worms are as painful as needles to the living; the deceased still have sentience, he avers. Another Rabbi claims that, in fact, the dead can feel their *own* pain, but not the pain of others. A third Rabbi disagrees. He suggests that the dead know things about pain and suffering, but feel nothing of them. Nothing we do in life affects the dead, though they know of it. And as proof, this last Rabbi recounts the following story, about which I would like now to think.

Once, on the eve of Rosh Hashana, the new year of judgment, and during a year of extreme drought, a certain pious man gave a dollar to the local beggar, and his wife scolded him for his charity. It must have been a fierce scold, because the man left his home and spent this holy night in the cemetery. I wonder why the Rabbis would send this man to the cemetery for comfort. I mean, for respite and consolation why wouldn't the man go to the *shul*,[15] or the *beit hamikdash?*[16] But no, the Rabbis send him to the cemetery! Perhaps in the cemetery nothing can further encumber this seemingly overly encumbered man. Here, in the cemetery, neither the responsibilities of prayer, nor study, nor human obligation are present, and the man can for a moment taste some freedom. In the silence of the cemetery, the man stands alone, unencumbered. If he would be around the dead for some comfort, perhaps it is because he thinks they will not trouble him. He has something yet to learn.

Or perhaps, on this eve of Rosh Hashana, the man comes to the cemetery, the place of final judgment, for some judgment. Deuteronomy 11:12 says that "The eyes of God . . . are always upon it [the land], from the beginning of the year to year's end." Judgment begins at the new year, and this story takes place on the eve of Rosh Hashana, the new year. Following a year of severe drought, and subsequent to a scolding by his wife for his charity, the man might wonder what judgment may be in store for him. On Rosh Hashana it is written who shall live and who shall die. Perhaps it is in the cemetery and not the shul where supplication may be best made; perhaps it is in the cemetery and not the *shul* or the *beit hamikdash* that learning may be best accomplished; perhaps it is in the cemetery and not the tavern that respite from his troubles might be best attained. Prospero's every third thought is his grave.

And remarkably, there, in the cemetery, the man heard two spirits talking. One spirit invites the other to "wander about the world and let us hear on this day of Judgement from behind the curtain what suffering is coming on the world." The spirit intends to seek out the misfortunes that will come to the world in the coming year, and only those will she report. On Rosh Hashana, the Jew prays to avert the severe decree, but cannot know it; only the spirits have access to this knowledge, but they cannot tell it. I wonder what it means to them. The Rabbis wonder about this: when the dead discover the misfortunes and suffering to which the world will be subject in the coming year, could it mean that the dead maintain feeling for the living? How might this obligate us, the living? Since our sufferings and misfortunes will only be known by the living ex post facto, so to speak, then we must keep on keeping on, trying as we might to minimize the inevitable. To this invitation the second spirit responded that because she had been buried not in a linen shroud but in matting reeds, she could not go out into the world, but that she would appreciate hearing the miserable news from beyond the curtain. I guess if we want the dead to maintain some access to our world, then we must bury them properly, that is, to treat them with honor and respect though we lay them in the ground.

So the other spirit went out and soon returned to her companion. "What have you heard from behind the curtain?" her stationary companion asked. And the first spirit said that she had heard that whoever sows his fields after the first rainfall, which would be about the middle of the month of Heshvan, will have his crop smitten by hail. The man, hearing this, decided not to sow till the second rainfall. Everyone's crop was ruined but his, and he made a great deal of money.

I wonder from whom the first spirit heard such news, and I must say, the story provides no clue. Clearly, however, if there is judgment at Rosh Hashana, certainly the living are not privy to it, unless, as in this story, they eavesdrop on the dead. To learn our fates, we must live our lives. Perhaps this is what Dylan (2006) means when he sings, "If I don't do anybody any harm, I might make it back home alive." Our lives are the judgment on our actions. I wonder upon what authority the man trusted the words of the spirit. Though neither spirit seems to be aware of the presence of the man, and does not know that they speak to the living, the living seem to know that the dead speak truths. Perhaps it is that the dead have no cause to prevaricate, and what they say must be truth. In this case, the dead seem to be no more than voyeurs, and the living mere accidental witnesses to the knowledge the dead gather from their visits to this world. I, myself, trust some of the dead, though not all of them; and I suspect that there are many of the dead who know things I would learn, but alas, I cannot yet hear them. Perhaps we teachers spend too little time in the cemeteries listening to the dead. Perhaps we should study more.

The next year the man again gave money to the same pauper on the eve of Rosh Hashana, and this time it was his son who berated him for his charity. Again, the man went to the cemetery for the evening, and heard the same two spirits talking. The first invited her companion to go out into the world and see what suffering would come in the new year. And again, the second spirit explained that she could not go out into the world because she had not been buried in her linen shroud, but that she would love to hear what suffering would come to the world during the new year. And again the second spirit went out and returned. This time she reported that "I heard that whoever sows after the later rain will have his crop smitten with blight." And so the man went and sowed his fields only after the first rain, with the result that everyone else's crop was blighted and his was not. The man's success was remarkable and singular. His wife was curious, and she asked him why he had planted in such manners, and he told her his experiences.

Shortly afterward, a quarrel arose between the wife of the man and the mother of one of the spirits, and the man's wife accused the latter of burying her daughter not in the traditional shroud, but in the less expensive and nonritual matting reeds. I suspect the mother of the dead woman was sorely shamed. Dead bodies must be properly buried.

And so, in the third year, the man on his own went to the cemetery on the eve of Rosh Hashana; of the dead's value, he had learned his lesson. And for the third time, the man heard the two spirits conversing. The first invited the second to come out beyond the curtain, and see what suffering would come this year into the world, and the other responded that she preferred to be left alone. Clearly, she said, since my mother has been accused of burying me in matting reeds instead of the traditional shroud (how else would the wife know that the daughter had been buried only in matting reeds), our conversation had been heard in the world, and now her mother was suffering. And the Rabbi concluded his story, "Wouldn't this show that the dead know of the suffering of others?" Alas, having been human they know of the suffering of others.

Finally, of course, whether the dead know of the suffering of others is of little matter: we suffer regardless of who knows of it. And their immortality, such as it is, has little to do with the world. Ah, the dead remain dead and incapable of intentional contact with the living. Unlike Elvira, in Noel Coward's play *Blithe Spirit*, the dead cannot return to haunt the living; the spirits talk as if the man is not present, and they have no idea that he overhears their conversation. The dead are curious but oblivious about the living. Perhaps that is one meaning of the story: we suffer, and we experience pain from that suffering alone. That the dead might share this pain is irrelevant, finally. We suffer nonetheless, and the sentience of their immortality is of little concern. As Dylan (1965) sings, "I've got nothing, Ma, to live up to." Our immortality resides in what we have here in the present, and that immortality depends on our responsibilities to the other. As Dylan sings, echoing the Unasaneh Tokef prayer for Rosh Hashana, "We live and we die, we know not why / But I'll be with you when the deal goes down." It was during his life that the father taught the sons to love learning; it was their own lives that led them from that learning and that finally returned them to it. When they forget their learning, it is not their father whom they forget. Rather, it is the other of whom their learning teaches: you must care for the stranger, the widow, and the orphan in your midst. Regardless of who else knows of it, our suffering is our own. Immortality may not be painless, but it is without pain.

What the dead know, however, can save us, if we can eavesdrop on their conversations. We do have some obligation to the dead, it would seem, to listen to their talk, though apparently, we must be in the right place and at the right time to benefit from their knowledge. And they apparently have no idea that they are being overheard. Nonetheless, the dead have no obligation to the living, and their immortality does not depend on what they have said but rather, on what we overhear. We must be attentive, I think. Dylan (2001) sings in "Sugar Baby," "You always got to be prepared, but you never know for what." We teachers prepare, but we never know for what. Of course, we must bury our dead properly—in shrouds and not in matting reeds—that they might continue to be present. Their immortality depends on us; they have none of their own.

And what if the dead do know of the suffering of others? Finally, they can do nothing about it; finally, it is we who must care in this life and not in the next. And this life exists in the present. It is all that we have of immortality.

As I write this I have entered my seventh decade. Every third thought is of the grade and the brave; every third thought will be my grave. My life is full.

Notes

Hakdamah

*A Hebrew word meaning "forepiece," as in moving forward.

1. During this Suzuki camp experience, there were scheduled support sessions for parents in which all we did was complain!

2. I have for a long time been fond of the distinction Winnicott makes between imagination and fantasy: the former I use to engage in creativity but the latter I use to avoid it by imagining engaging in imagination. With the former I write the Pulitzer Prize-winning text, and with the latter I write the acceptance speech for the prize for the book I haven't written while I fantasized the acceptance speech.

3. According to legend, until someone stepped into the Sea, the waters would not part. The people were reticent to take the first step, waiting for God to perform the miracle. Only when Nachshon stepped into the Sea of Reeds did the waters part.

4. *Mitzrayim*, the Hebrew word for Egypt, means the narrow place, alas, where even thoughts are constrained.

5. *Midrash* is a story explaining the text. It is a hermeneutical method of dealing with text.

6. This is, perhaps, the first hint that it is by human action that finally the people might be sustained.

7. The word manna seems to derive from the Hebrew: when the people see the substance on the ground they wonder "what is this?" The Hebrew for this is *man-hu*.

8. I recall that Lionel Trilling was the first Jew promoted to tenure at Columbia University in 1947!

9. The Mishnah is the body of law about which the Rabbis' conversation engages. In the course of discussion, the Rabbis introduce law, folklore, magic, legend, and stories.

Chapter 1   Why Be a Teacher

1. Classroom teachers regularly refer to this daily work as "life in the trenches." I intend to address this phrase in part 2 of this text.

2. Even as I wrote those words I thought of Steve Earle's anthemic song, "I Ain't Ever Satisfied." Somehow, in Eisner's list there is the hint that dissatisfaction is anathema to the profession, but for Earle, dissatisfaction is endemic to active living. Before his death, Zusya (in Buber, 1951, 251) said, "In the coming world, they will not ask me: 'Why were you not Moses?' They will ask me: 'Why were you not Zusya?'" Life is all about incompletion.

3. This last aspect of teaching smacks to me as a form of immortality, explanation two, but I will not quibble.

4. Barnes and Noble had prominently displayed a black cover edition of Hitler's *Mein Kampf,* and a white cover edition of David Horowitz's book *The Professors: America's One Hundred Most Dangerous Professors.* The former is well known; the latter is the latest attempt to institute the blacklist, the witch hunt, and the Alien and Sedition Acts.

5. Though, to look at the setting sun in another way, I offer this: Thoreau (2001, 255) writes: "The sun sets on some retired meadow, where no house is visible, with all the glory and splendor that it lavishes on cities, and, perchance, as it has never set before,—We walked in so pure and bright a light, gilding the withered grass and leaves, so softly and serenely bright, I thought I never bathed in such a golden flood, without a ripple or a murmur to it." Methinks the Grateful Dead read Thoreau as well (listen to "Ripple," on *American Beauty*).

6. An article in the *Washington Post* (http://www.washingtonpost.com/wp-dyn/content/article/2006/05/08/AR2006050801344.html?sub=AR) reports that 50 percent of teachers leave the profession within five years.

7. Indeed, I think some did learn some American history. I met one boy several years after who told me that one song he learned in my class helped him answer a question on a multiple choice test he took in high school. I was not comforted.

8. Yes, I acknowledge, I have earned tenure, and can act with less caution.

9. This is a metaphorical statement. Depending on the type of schedule a school maintains, classes might run anywhere from forty-five minutes to ninety minutes.

Chapter 2   A Pedagogue for Two Teachers

1. The elector princes of the Holy Roman Empire were the members of the electoral college of the Holy Roman Empire, having the function of electing the Emperors of Germany. From the thirteenth century on, there were seven electors altogether, three spiritual and four lay ones. The Count Palatine of the Rhine was the Elector Palatine.

2. It is this role of the teacher that has led to the proliferation of methods and classroom management courses, and to the prevalent ideology that floats about that in schools of education that "it is not what you taught be but how you made me feel." The title of Herbert Kohl's book (1994) derives from a statement made by a student: "I won't learn from you!"

3. Unless, of course, we acknowledge that maintaining employment is good for the teacher.

4. In February, 1671, Spinoza (1955, 363) wrote to I.I (probably Jarig Jellis, a merchant in Amsterdam, begging him "if possible, [to] stop the printing in Dutch of his *Tractatus-Theolgicus*, to prevent that book from being placed under an interdict."

5. There are several explanations for Spinoza's choice of method and structure. See Smith (2003, Chapter 1).

6. Of course, there were many who hated the book though they did not actually read it. Their opinions derived from rumor and public report. Today academic publishers break even if two hundred copies of a book are sold, and may not really care what the critical opinion is beyond the sale of those first two hundred. Alas, I don't know that I have been always able to balance the accounts of my publishers.

7. See an exploration of the causes of the *cherem* in *Spinoza's Heresy* (Nadler, 2001).

8. Of course, it is not uninteresting that Dewey, too, has been the target of vilification, and his work is still today blamed for all varieties of America's problems. I remember once reading that it was the result of Dewey's pedagogical theories that America's communists prospered. Somewhere, I am sure, somebody has blamed Dewey for sweetened breakfast cereals.

9. I am incensed by Hirsch's insinuation that teachers do not *already* teach content-knowledge. Indeed, it is just this charge that is the target of the standards and standardized tests.

Chapter 3   On Finding Lost Objects

1. The Mishnah is the body of law redacted in about 200 CE by Judah Ha-Nasi. The Gemara is the body of commentary concerning that law. The Gemara was edited from notes of the Rabbis' discussions between 200 and 600 CE and has been continually added to ever since.

2. Halakah traditionally refers to "the law." It literally means "the way one should walk."

3. I suppose one could argue that the misogyny of the Rabbis write women out of the conversation but the movement of their thinking often reveals the Rabbis struggling to overcome their own misogyny. I think they recognize their contradictions, and work to resolve them in a manner that seeks to enlarge their culturally imposed bounds. This is the argument of Judith Hauptman in her book *Rereading the Rabbis* (1998, 4). There she argues that though the Rabbis clearly upheld the patriarchy "as the preordained mode of social organization...they began to introduce numerous, significant, and occasionally bold corrective measures to ameliorate the lot of women...From their own perspective the rabbis were seeking to close the gap that had developed over time between more enlightened social thinking and women's more subordinate status as defined by the received texts, biblical and rabbinic, without openly opposing such texts." Daniel Boyarin argues in *Carnal Israel* (1993, 242) that Rabbinic texts may be read to recover "feminist" voices. "The texts when read in the way that I

am proposing to read them do not only reflect a dissident proto-feminist voice within Classical Judaism; they constitute such a voice." Finally, though the Rabbis hold to a belief in the world to come (see Chapter 10), I am reading this statement literally to refer to a future in this world.

4. The Gemara is the multigenerational redacted commentary on the Mishnaic law. It was formally published along with the Mishnah in about 600 CE. Together, Mishnah and Gemara comprise the Talmud.

5. Zohama listron is a utensil with a spoon on one side and a fork on the other. Apparently, teaching even one thing from a text qualifies someone as a teacher.

CHAPTER 5   IT MUST SUFFICE

1. A colleague was once told by a concerned parent that since he was, as a teacher, also a magician, he should be competent to deal with her rebellious child.

2. Ironic it is that I tutored the tutor. He asked if I could help him prepare for the Graduate Record Exams.

CHAPTER 6   STUDY AND BENEVOLENCE

1. This book won the Outstanding Book Award in Division B Curriculum Studies of the American Educational Research Association, but my friend, David, reminds me to repent for commission of the sins of commerce.

2. I think this critique predated Bennett's admission of his multimillion dollar loss at the gambling tables.

3. Of course, by "repair" we do not mean physical repair of which there is great need. For this sort of effort, we regret to inform you, we have insufficient funds.

4. Of course, this statistic disproves Bennett's harsh critique.

5. The term *epikoros* derives from the Greek and refers to a follower of Epicurus, or one who lives a dissolute and licentious life. It might also refer to a heretic. Hence, the term is used to refer to someone who speaks disparagingly of the Torah and its students.

6. Manasseh reigned in Judah from 687 to 642 BCE, and overturning the reforms instigated by his father, Hezekiah, reinstated pagan worship in the Temple.

7. Jeraboam reigned from 922 to 901 BCE, and was the first king of the break-away ten tribes. He built a northern temple with calf-like idols.

8. Ahab, under the influence of his non-Jewish wife, Jezebel, built a temple to Baal in Samaria.

9. Baalam was hired by Balak before he led his attack, to curse the encamped Israelites' nation.

10. Doeg was an Edomite who betrayed David.

11. Ahitophel was an advisor to King David whom he betrayed to side with Absalom.

12. Gehazi was Elisha's servant who was overcome by avarice and received payment in Prophet Elijah's name from Naaman.
13. Hebrew is a consonantal language and written without vowels; words can be variously identified as a result.

CHAPTER 7 INTIMATIONS OF IMMORTALITY: AN ODE OF SORTS

1. Alas, I have lost the program sheet, and cannot find the poem in question. Alas, perhaps I have misremembered.
2. Holden Caulfield commits a similar mishearing; he understands Burns' song to say "when a body catch a body, comin' through the rye" when in actuality the line is "when a body *meet* a body comin' through the rye." We hear what we want to hear. I once read a lovely essay on Holden's mishearing.
3. With my most sincere apologies to William Wordsworth.
4. These beans are purchased at a variety of specialty outlets, Starbucks being, of course, a significant source, but J&S, a privately owned coffee emporium, has become my newest and favorite cup of choice.
5. Unlike Thoreau, I cannot bathe this early, especially not in the cold ponds of Menomonie. Well, or anywhere else for that matter. Part of my ritual behavior is an easing, rather than a plunging into the day!
6. "We call a thing bad when it is the cause of pain" (IV.viii), "when it diminishes or checks our power of action . . . in so far as it can diminish or check our power of action, it is contrary to our nature" (Spinoza, 1955, 206–207).
7. The mind is action perceived through the attribute of thought.
8. *Berakhot* means "blessings."
9. See Goce Smilevlski's *Conversation with Spinoza: A Cobweb Novel* (2006).
10. Thoreau died in 1862, having begun to work on the material for this essay in 1857. In his final months, he finished "Autumnal Hints."
11. See how the Rabbis often in their discussions redefine words to suit their meaning, and see Thoreau's lovely analysis in "Walking" of the derivation of the word "to saunter."
12. Dylan sings, "You can always come back, but you can't come back all the way." Hmnn.
13. Serah is the daughter of Asher, and the story narrates that it was she who told Jacob that Joseph was still alive. For this Jacob blessed her and she was granted immortal life. It is also said that it was Serah who told Moses where Joseph's bones were buried when the Israelites left Egypt.
14. And perhaps, the story narrates, we must act in eternity as if we have not left this world; if the dead are sentient and feel pain, the dead still show concern for this world. In this way, perhaps, we achieve immortality.
15. The House of Prayer.
16. The House of Study.

References

Bauman, Z. (1993). *Postmodern Ethics.* Oxford: Blackwell.

Berliner, D. C., and Biddle, B. J. (1995). *The Manufactured Crisis.* New York: Addison-Wesley.

Block, A. A. (1999). The Book That Changed My Life, or What I Didn't Read on My Summer Vacation? *Journal of Curriculum Theorizing. 15:4.*

———. (1998). Curriculum as Affichiste. In W. F. Pinar (Ed.), *New Curriculum Identities.* New York: Garland.

———. (2004). *Talmud, Curriculum and the Practical: Joseph Schwb and the Rabbis.* New York: Peterlang.

———. (2007). *Pedagogy, Religion and Practice: Reflections on Teaching and Ethics.* New York: Palgrave Macmillan.

Block, A. A., and Reynolds, W. (1994). Curriculum as Making Do. In R. A. Martusewicz and W. M. Reynolds (Eds.), *Inside Out: Contemporary Critical Perspectives in Education* (pp. 209–221). New York: St. Martin's.

Bollas, C. (1992). *Being a Character.* New York: Hill and Wang.

Boyarin, D. (1993). *Carnal Israel.* Berkeley: University of California Press.

Bracey, G. W. (2003). *On the Death of Childhood and the Destruction of Public Schools: The Follow of Today's Education Policies and Practices.* Portsmouth, NH: Heinemann.

———. (2005). America's Groundhog Day. Retrieved November 11, 2005, from http://fotps.org/media/article-bracey.pdf.

———. (2006). *Reading Educational Research: How to Avoid Getting Statistically Snookered.* Portsmouth, NH: Heinemann.

———. (1975). *Tales of the Hasidim.* New York: Schocken Press.

Darling-Hammond, L. (2006, May/June). Constructing 21st-Century Teacher Education. *Journal of Teacher Education, 57*(3), 300–314.

Deleuze, G. (1988). *Spinoza: A Practical Philosophy* (R. Hurley, Trans.). San Francisco: City Lights.

Dewey, J. (1916/1966). *Democracy and Education.* New York: Free Press.

———. (1964). *John Dewey on Education* (R. D. Archambault, Ed.). Chicago: University of Chicago Press.

———. (1974). My Pedagogic Creed. In R. D. Archambault (Ed.), *John Dewey on Education.* Chicago, Illinois: University of Chicago Press.

Doctorow, E. (1971). *The Book of Daniel.* New York: New American Library.

Dylan, B. Idiot Wind. In *Blood on the Tracks*. New York: Columbia Records (Original work published 1974).

———. It's Alright Ma (I'm Only Bleeding). In *Bringin' It All Back Home*. New York: Special Rider Music (Original work published 1965).

———. Mississippi. In *Love & Theft*. New York: Columbia Records (Original work published 2001).

———. Sugar Baby. In *Love & Theft*. New York: Special Rider Music (Original work published 2001).

———. Tryin' to Get to Heaven. In *Time Out of Mind*. New York: Special Rider Music (Original work published 1997).

———. (1997). Not Dark Yet. In *Time Out of Mind*. New York: Columbia Records.

———. (2006). Nettie Moore. In *Modern Times*. New York: Special Rider Music.

Eisner, E. (2006, March). The Satisfactions of Teaching. *Educational Leadership, 63*(6), 44–46.

Emerson, E. W. (1999). *Henry Thoreau as Remembered by a Young Friend*. Mineola, NY: Dover.

Fiedler, L. (1991). *Fiedler on the Roof*. Boston: David Godine.

Gilman, S. (1991). *The Jew's Body*. New York: Routledge.

Gillman, N. (1999). *The Death of Death: Resurrection and Immortality in Jewish Thought*. Wootstock, Vermont: Jewish Lights.

Giroux, H. A. (1992). *Border Crossings: Cultural Workers and the Politics of Education*. New York: Routledge.

Halbertal, M., and Halbertal, T. H. (1998). The Yeshivah. In A. O. Rorty (Ed.), *Philosophers on Education*. New York: Routledge.

Halivni, D. W. (1997). *Revelation Restored: Divine Writ and Critical Responses*. Boulder, CO: Westview Press.

Hampshire, S. (2005). *Spinoza and Spinozism*. New York: Oxford University Press.

Handelman, S. A. (1982). *The Slayers of Moses*. New York: State University of New York Press.

Hauptman, J. (1998). *Rereading the Rabbis*. Boulder, CO: Westview Press.

Heschel, A. J. (1951). *The Sabbath*. New York: Farrar, Straus & Giroux.

———. (1959). *Between God and Man: An Interpretation of Judaism*. New York: Free Press.

———. (1962). *The Prophets*. New York: Harper/Torchbooks.

———. (2006). *Heavenly Torah* (G. Tucker, Ed. and Trans.). New York: Continuum.

Hirsch, E. (2006, Spring). Building Knowledge. *American Educator, 30*(18–21), 28–29, 50.

Howarth, W. (1982). *The Book of Concord*. New York: Penguin Books.

Hunter, R., and Garcia, J. Till the Morning Comes. In *American Beauty*. San Francisco: IceNine. (Original work published 1970).

Irving, W. (1961). *The Sketch Book*. New York: Signet Classics.

James, W. (1892/1961). *Psychology: The Briefer Course* (G. Allport, Ed.). Notre Dame, IN: University of Notre Dame.

———. (1962). *Talks to Teachers on Psychology and to Students on Some of Life's Ideals.* Mineola, NY: Dover.

Kayser, R., and Einstein, A. (1946). *Spinoza: Portrait of a Spiritual Hero.* New York: Philosophical Library.

Kliebard, H. (1999). *Schooled to Work: Vocationalism and the American Curriculum, 1876–1946.* New York: Teachers College Press.

Kohl, H. (1994). *"I Won't Learn from You," and Other Thoughts on Creative Maladjustment.* New York: New Press.

Kolitz, Z. (1995). *Yossel Rakover Speaks to God.* Hoboken, NJ: Ktav.

Kozol, J. (1992). *Savage Inequalities: Children in America's Schools.* New York: Harper Perennial.

———. (2005). *The Shame of the Nation.* New York: Crown.

Lambert, L. (2006, May 9). Half of Teachers Quit in 5 Years. *Washington Post.* Retrieved from http://www.washingtonpost.com/wp-dyn/content/article/2006/05/08/AR200605080134.

Levinas, E. (1990). *Nine Talmudic Readings* (A. Aronowicz, Trans.). Bloomington: Indiana University Press.

———. (1994). *In the time of Nations* (M. B. Smith, Trans.). Bloomington: Indiana University Press.

———. (1999). *New Talmudic Readings* (R. Cohen, Trans.). Pittsburgh, PA: Duquesne University Press.

Maimonides, M. (1963). *The Guide of the Perplexed* (S. Pines, Trans.). Chicago: University of Chicago Press.

Mandel, S. (2006, March). What New Teachers Really Need. *Educational Leadership, 63*(6), 66–69.

Miller, P. (1956/1964). *Errand into the Wilderness.* New York: Harper/Torchbooks.

Milner, M. (1950). *On Not Being Able to Paint.* London: Heinemann Educational Books Ltd.

———. (1987). *The Suppressed Madness of Sane Men.* London: Tavistock.

Morgan, M. (2005). Levinas and Judaism. In J. Bloechl and J. L. Kosky (Eds.), *Levinas Studies: An Annual Review* (pp. 1–17). Pittsburgh, PA: Duquesne University Press.

Nadler, S. (1999). *Spinoza, a Life.* New York: Cambridge University Press.

———. (2001). *Spinoza's Heresy: Immortality and the Jewish Mind.* New York: Oxford University Press.

———. (2006). *Spinoza's Ethics: An Introduction.* New York: Cambridge University Press.

Null, J. W. (2007). *Peerless Educator: The Life and Work of Isaac Leon Kandel.* New York: Peterlang.

Phillips, A. (1993). *On Kissing, Tickling, and Being Bored.* Cambridge, MA: Harvard University Press.

———. (1994). *On Flirtation.* Cambridge, MA: Harvard University Press.

Pinar, W. F. (2004). *What Is Curriculum Theory?* Hillsdale, NJ: Lawrence Erllbaum.

Pinar, W. F., and Grumet, M. (1976). *Toward a Poor Curriculum.* Kendall Hunt: Dubuque, Iowa.

Ravitch, D. (1994). *National Standards in American Education: A Citizen's Guide.* Washington, DC: Brookings.

———. (1995). *National Standards in American Education.* Washington, DC: Brookings Institution Press.

Sacks, O. (1985). *The Man Who Mistook His Wife for "A Hat and Other Clinical Tales."* New York: Summit Books.

———. (1969). *College Curriculum and student Protest.* Chicago: University of Chicago Press.

Schwab, J. J. (1971). The Practical: Arts of the Eclectic. *School Review, 79,* 493–542.

Scruton, R. (1999). *Spinoza.* New York: Routledge.

Shaffer, P. (1973). *Equus.* New York: Penguin Books.

Smilevski, G. (2006). *Conversation with Spinoza* (F. Korzenski, Trans.). Evanston, IL: Northwestern University Press.

Smith, S. B. (1997). *Spinoza, Liberalism, and the Question of Jewish Identity.* New Haven, CT: Yale University Press.

———. (2003). *Spinoza's Book of Life: Freedom and Redemption in the Ethics.* New Haven, CT: Yale University Press.

Spinoza, B. d. (1955). *On the Improvement of the Understanding; The Ethics; Correspondence* (R. Elwes, Trans.). Mineola, NY: Dover.

———. (2005). *A Theologico-Political Treatise and a Political Treatise* (R. Elwes, Trans.). New York: Cosimo Books.

Thoreau, H. D. (1834/1961). *A Week on the Concord and Merrimack Rivers.* New York: Crowell.

———. (1962). *The Journal of Henry D. Thoreau.* Mineola, NY: Dover.

———. (1970). *The Annotated Walden* (P. Van Doren Stern, Ed.). New York: Marboro Books.

———. (1980). Walking. In *The Natural History Essays.* Salt Lake City, UT: Peregrine Smith.

———. (1991). *Uncommon Learning: Henry David Thoreau on Education* (M. Bickman, Ed.). Boston: Mariner Original.

———. (2001). *Thoreau: Collected Essays and Poems.* New York: Penguin Books.

Vonnetgut, K. (1969). *Slaughterhouse Five, or the Children's Crusade.* New York: A Delta Book.

Wordsworth, W. (1967). Ode: Intimations of Immortality. In D. Perkins (Ed.), *English Romantic Writers.* New York: Harcourt, Brace, and World.

Index